TON

D1376158

Marine Dinoflagellates of the British Isles

John D. Dodge

Professor of Botany, Royal Holloway College
University of London

with the assistance of
Barbara Hart-Jones

WITHDRAWN

POLYTECHNIC LIBRARY
WOLVERHAMPTON

ACC. No.	CLASS
555438	579,
CONTROL	87
	DOD
DATE	SITE
-2 AUG. 1982	R S

D22

LONDON HER MAJESTY'S STATIONERY OFFICE

WP 0555438 1

593.18/DOD

©Crown Copyright 1982

First Published 1982

ACKNOWLEDGEMENT

The publication of this monograph was made possible by the generous assistance of the Ministry of Agriculture, Fisheries and Food, Directorate of Fisheries Research and was typeset by the MAFF Specialist Typing Unit.

The method of typesetting used for this book unfortunately prohibited the use of accents and other diacritical marks.

ISBN 0 11 241196 7*

CONTENTS

Preface

In 1967 I was asked by the late Roger Butcher to write a volume on the Dinophyceae for the Fisheries Investigations Series IV of Monographs which he had commenced under the general title "An Introductory Account of the Smaller Algae of British Coastal Waters". Sadly the series has now ceased although valuable works on the green flagellates and the Cryptophyceae (Butcher) and the Bacillariophyceae (Hendey) were published.

With the financial assistance of the Natural Environment Research Council from 1969–1978 it was possible for me to have full-time assistance with the collection of literature and the examination of samples. At first this task was carried out by June Carslake (1969–72) but for most of the period and including much of the drafting of the manuscript I was capably assisted by Barbara Hart-Jones. The drawings, which illustrate almost every species listed, are my own work and where possible have been made from fresh observations and photographs. The scanning micrographs in plates I–VIII were taken either at Birkbeck College by Grahme Lawes or at Royal Holloway College by Heather Hermes.

The NERC support also enabled us to visit many sites around the British Isles to collect samples and to examine the delicate naked species. We were also able to arrange for samples to be sent to us by marine laboratories. The great majority of samples which we have examined were kindly provided by the various workers who are listed on p. 271. We received much help from the various MAFF laboratories. At the start of the project I was greatly helped by Mrs D. Hepper (D. Ballantine), who generously allowed me to use her extensive card index of dinoflagellate literature and species descriptions. To all who have assisted in any way – I wish to express my deep gratitude. In particular I am grateful to Dr M Parke for first interesting me in dinoflagellates and for her encouragement at various times. Special thanks must go to the staffs of the Botany Departments at Birkbeck College and latterly Royal Holloway College for excellent photographic and secretarial assistance.

I would make a final plea for indulgence. The rate of taxonomic advance is so great, names are changing with such rapidity that by the time this appears in print it will sadly be out of date. However, I hope that it will in some small way help with the advancement of our knowledge of the fascinating dinoflagellates which occur in the seas around the British Isles.

May 1980   John D. Dodge

INTRODUCTION

Apart from "The Dinoflagellates of Northern Seas" by Lebour (1925) there has been no comprehensive study of the dinoflagellates found around the British Isles. In fact, considering their potential importance both in the marine food webs and as organisms which can cause sea food to become toxic to man, very little work has so far been carried out. Notable pioneers apart from Lebour were Catherine Herdman who showed something of the range of dinoflagellates living on sandy shores and Mary Parke who first succeeded in isolating and culturing a range of neritic dinoflagellates and was responsible (with D. Ballantine) for collecting together the first check-list of dinoflagellates which have been recorded from around the British Isles and adjacent waters (3rd Edition: Parke and Dodge, In: Parke & Dixon, 1976).

In the early years of the century there was an attempt to survey the dinoflagellates found in areas important for fishing such as the English Channel and North Sea. Some general surveys (e.g.: Gough, 1908) were published together with distribution maps for the Peridinia (Paulsen, 1911) and Ceratia (Jorgensen, 1912). Much more recently, in connection with the present study, further ecological surveys have been carried out in areas of the North Sea (Dodge & Hart-Jones, 1974; 1977; Dodge, 1977). Of considerable interest in relation to the formation of red tides of possibly toxic dinoflagellates have been the recent studies of *Gyrodinium aureolum* blooms associated with several oceanographic frontal systems around the British Isles (Holligan & Harbour, 1977; Pingree *et al.*, 1976; 1977; 1978; Holligan, Maddock & Dodge, 1980).

Scope of the Work

The object of this work is to provide a book which can be used by fisheries workers and others for the identification of the more common dinoflagellates to be found in the open seas, coastal waters, estuaries and shores around the British Isles. Many of these organisms are also found in other parts of the world and some indication of their distribution is given following the descriptions. The normal limit of the area covered will be seen by reference to the maps 1 & 2, (p. 5). It comprises the North Sea, English Channel, Irish Sea, the Atlantic out to about 10°W and the area north of Scotland up to almost 61°N. Weather Station India, in the North Atlantic (59°N, 19°W), was studied extensively.

In deciding which species to include we have firstly included the 200 or so species which have been seen during the course of the survey. Here, there is a preponderance of thecate organisms because so much of the work was carried out on fixed material. Other species fairly reliably reported from around the British Isles have also been included as also some found off Continental Europe but which should occur here. Apart from the genus *Dissodinium*, which has a clearly recognisable free-living stage, no parasitic species have been included. The total species described or briefly mentioned comes to about 250 representing 58 genera compared to the 440 species representing 63 genera in the latest check-list (Parke & Dixon, 1976). The biggest discrepancy is in the main genera of naked organisms, *Amphidinium*,

Gymnodinium and *Gyrodinium*. It is clear that the literature contains many ill-described taxa which are the result of bad fixation or shape-change during observation. This is an area where far more work is needed both by rapid observation following collection of samples, by development of better fixation methods and by isolating organisms into culture for more careful study.

Structure of dinoflagellates

Most dinoflagellates are single-celled organisms which contain all the structures necessary for life within the compass of one variously shaped cell. There is a fairly large nucleus with chromosomes which, in most species, are permanently condensed and can be readily seen by phase-contrast microscope or after staining. One or more chloroplasts of a yellow-brown colour may be present and these owe their particular hue to a combination of the pigments chlorophyll a, chlorophyll c, β carotene and either peridinin or fucoxanthin (see Jeffrey *et al*. 1975). The ultrastructure of the chloroplasts is distinctive (Dodge, 1975). Non-photosynthetic dinoflagellates often have a pale pinkish coloration. Most species have trichocysts which eject from the cell under adverse conditions and many possess a pair of elaborate organelles called pusules which are associated with the flagellar pores and probably function in osmo-regulation (Dodge, 1972). Much more complex ejectile bodies or nematocysts are found in *Polykrikos* and some members of the family Warnowiaceae (Greuet, 1971). The latter also have highly complex ocelli (Greuet, 1978) which contrast with the relatively simple eyespot found in a few marine dinoflagellates (Dodge & Crawford, 1969).

One characteristic of the Dinophyceae is the presence of two flagella, which may be apically or laterally inserted, and which are dissimiliar. One is of fairly normal construction and may have delicate hairs near the tip, the second is unique with a helical or ribbon-like construction and a unilateral array of long hairs (Leadbeater & Dodge, 1967; Taylor, 1975a). The cell covering, variously termed the theca or amphiesma, is a complicated construction consisting of an outer membrane with underlying membrane vesicles. In the 'naked' genera these vesicles remain virtually empty but in armoured genera each contains a thecal plate of varying thickness, shape and ornamentation (Dodge & Crawford, 1970). The characteristics of the theca are of prime importance in the taxonomy of the group (see Fig 1).

Reproduction

Reproduction is generally by binary fission which follows nuclear division. However, the parasitic organisms may have a much more complex division associated with their usually more elaborate life cycle. Sexual reproduction was for long an uncertain matter in these organisms but recently the careful work of Von Stosch (1973) and the need for sexual reproduction to facilitate genetic studies (Beam & Himes, 1977) has shown that pairing of motile gametes can give rise to motile zygotes and eventually to resting hypnozygotes or cysts.

Cysts

Many of the resistant stages are similar to fossil cysts known from Jurassic to present day deposits. The work of Wall & Dale (1968a; 1968b) has shown the connection between many of these cysts and the thecate stages. Where the correlation is clear this has been detailed in the descriptions which follow. Of considerable importance is recent work on the cysts of toxic dinoflagellates such as *Gonyaulax tamarensis* (Anderson & Wall, 1978; Turpin *et al.*, 1978) these appear to be both toxic themselves and to provide the overwintering stage which allows a 'red-tide' to form rapidly when conditions become suitable in the next spring or summer.

Life form

The majority of dinoflagellates are free-living but a small number of species with considerable importance, particularly in the ecology of coral reefs, are found as symbionts. A number of species are parasites on copepod eggs, on and within copepods, on appendicularians etc. A few species mainly exist as coccoid cells which have a motile stage and there have been a few reports of a filamentous organism (*Dinothrix*) which has gymnodinioid zoospores. The present work concentrates on the free-living and symbiotic forms and only deals with *Dissodinium* amongst the parasitic genera.

Nutrition

Apart from the specialised nutritional types mentioned above dinoflagellates are almost equally divided between those which are chloroplast-containing, and therefore can be autotrophic, and those which lack chloroplasts and must be heterotrophic. A small number are definitely phagotrophic and it may be that this form of nutrition is more widely practised than is immediately apparent.

Red Tides

Perhaps the feature of dinoflagellates which brings them mostly into the news is their ability to form red tides which are in effect high concentrations of cells in limited areas of the sea. Many factors have been put forward to account for these phenomena which have been known at least since Darwin's voyage round the world in 1831. The temperature must be high enough, adequate nutrients must be available and, in the case of *Gonyaulax tamarensis,* it is now known that removal of inhibitors (in this case Copper) by natural chelating agents can allow bloom conditions to develop (Anderson & Morel, 1978). Most red-tides are coastal phenomena and they usually depend on temperature or salinity stratifying effects which increase the concentration of cells in a small area such that there may be as many as 75 million per litre of *Gymnodium breve* in the Florida red tide (Steidinger, 1973).

Toxicity

The main problem with red tides is that they generally consist of species which can produce toxin. There are two types of toxin, that secreted by the flagellates (such as *G. breve*) and mainly active against fish, and an endotoxin characteristic of the genus *Gonyaulax* which is only obtained from the dinoflagellate by filter-feeding animals. These can then concentrate the toxin and present it in a form which can affect man and other vertebrates. Further information on these species and the intensely toxic poisons which they produce can be found in the reports of the symposia which have been held on this topic (Lo Cicero Ed. 1975; Taylor & Seliger Ed. 1979).

Bioluminescence

Another phenomenon often associated with red tides is bioluminescence. A number of dinoflagellates have the ability to emit flashes of light which are produced by a luciferin/luciferase system as in the firefly. In some species the system is rhythmic, in others it requires stimulation to set it off. In tropical waters genera such as *Pyrocystis* and *Pyrodinium* can provide a striking night spectacle. Around the British Isles *Noctiluca* often appears in sufficient numbers in the summer to illuminate the sea at night and a number of *Protoperidinium* species are also known to perform (Tett, 1971).

Ecology

In simple terms the marine dinoflagellates can be divided into two main ecological types, oceanic and neritic. The oceanic species are often large (e.g. many Ceratia) and can grow at very low nutrient levels. They are not adapted to any great change in temperature or salinity and very few have been cultured in the laboratory. The neritic forms can be further subdivided. Firstly, there are open-water neritic species which exist in the sea adjacent to land. These may form blooms under suitable conditions and they are subject to fairly substantial changes in the condition of the sea water. Some only appear to thrive in the brackish water of estuaries. A few species are found in the semi-permanent pools of salt marshes where great extremes of salinity and temperature are experienced. Some of these species also occur in high-tide pools on rocky shores. Finally, a number of dinoflagellates occur in a littoral situation living between sand-grains on damp beaches, often forming a brown crust when the water is low but retreating amongst the sand when the tide covers the beach. Such species are usually flattened. The majority of marine dinoflagellates in culture collections are those of neritic situations.

Distribution

The distribution of dinoflagellates around the British Isles depends on the factors outlined above but in addition there are the contrasting influences of the Gulf-stream to the west and the Continent of Europe to the east. Thus the west of the British Isles receives a constant input of warm water, with high salinity, containing oceanic Atlantic species. The east coast by way of contrast is extremely cold in winter and in the summer is cool to the north and very warm to the south. This is clearly reflected in some of the distribution patterns. Maps giving the outline results of a ten-year survey of the distribution of dinoflagellates around the British Isles in the area covered by Map 1 have been published in 'A Provisional Atlas of the Marine Dinoflagellates of the British Isles' (Dodge, 1981).

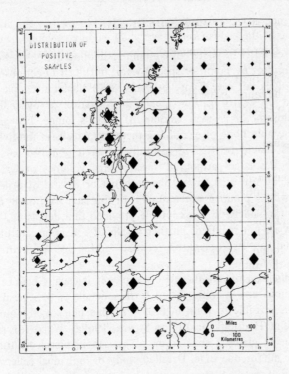

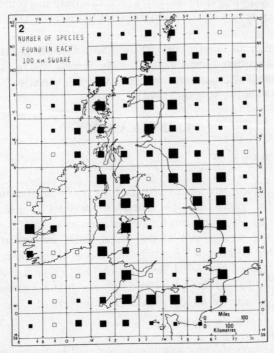

Map 1   Samples per 100 km square   Map 2   Species per 100 km square

♦ > 50   ■ > 60

♦ 10–49   ■ 40–59

• < 10   ▪ 20–39

METHODS OF STUDY

Samples of sea water suspected of containing dinoflagellates may be collected by simple dipping, by pump or special water bottle. If direct observation by microscopy does not reveal any, or many, cells then the sample should be concentrated by centrifugation, membrane filtration or by sedimentation after the addition of a small amount of Lugol's iodine. For naked dino-flagellates the only hope of examining them properly is in the live state and thus samples should be treated carefully when it is hoped to study such species. For the larger armoured species large volumes of water can be concentrated by use of a towed plankton net or by use of a net mesh in association with a pumped water supply. In this case fixation in neutralised or buffered formaldehyde is acceptable provided that details of flagella and cell contents are not required. Further details of collection and preservation will be found in "Phytoplankton Manual" edited by A. Sournia (1978).

For examination of dinoflagellate samples it is usually best to first look at these without any staining. When further details of the thecal plates are required it is possible to stain in 0.25% Trypan blue or, for more critical study, in Iodine/Hydriodic acid (Von Stosch, 1969a). Where it is required to dissociate thecal plates a small drop of household bleach solution can be added to the microscope slide. Detailed study of individual cells which have been picked out, washed, metal coated, and examined in a scanning electron microscope can be very informative (see plates I—VIII).

For the study of dinoflagellate cysts from sediments special palaeonto-logical techniques have been developed. These involve sieving after acid solu-tion of calcareous siliceous and non-resistant organic remains. Details of this technique are found in Sargeant (1974).

CLASSIFICATION

Basis of the classification

Naked dinoflagellates are mainly classified on the basis of their shape, the position of the girdle and the general proportions of the cell. These characteristics are also important for armoured dinoflagellates but in addition the number and arrangement of the thecal plates are important. In the Peridiniales the method of describing the thecal plates (as used here) is based on one devised by Kofoid. Recently Taylor (1979) has proposed a new scheme intended to facilitate comparison between different genera and Eaton (1980) another scheme which should aid comparisons with fossil dino-flagellates.

The classification of living organisms is governed by the codes of nomen-clature which are slightly different for the various groups. Here, the dino-flagellates cause great problems because some workers regard them as animals and follow the International Code of Zoological Nomenclature whilst others regard them as plants and follow the International Code of Botanical Nomen-clature. In reality dinoflagellates as a group are neither animals nor plants and there is a clear need for an agreed approach for such intermediate groups.

A further problem comes from the increasing importance of dinoflagellate cysts which, if described by geologists, may have generic and specific names quite distinct from the motile stages to which they relate.

Many schemes have been proposed for the classification of dinoflagellates. The major division is normally termed the phylum or division PYRROPHYTA or DINOPHYTA in botanical terminology, or the order DINOFLAGELLIDA or DINOFLAGELLATA of the class Mastigophorea in protozoological classification. Within the division or order traditionally dinoflagellates have been divided into two groups based on the position of insertion of the flagella. These have been given various names: Adiniferidea, Diniferidea (Lebour); Desmokontae, Dinophyceae (Schiller); Adinides, Diniferides (Deflandre); Desmophycae, Dinophyceae (Loeblich, 1970). It has been suggested (Dodge & Bibby, 1973) that there is no basic distinction in terms of cell structure, ultrastructure or of pigmentation etc. between the two groups and so in this work we follow Parke & Dodge (1976) in including them both within the class DINOPHYCEAE.

The class is subdivided into a number of orders of which the following are included in this work:-

Prorocentrales, in which the flagella are inserted anteriorly and there are only two main thecal plates.

Dinophysiales, where the flagella are inserted at one side near the anterior end of the cell but a distinct transverse groove or girdle is present. There are thecal plates.

Gymnodiniales, in which there are no substantial thecal plates, i.e. the cells are naked. Various shapes and forms are found. It is open to question whether the Lophodiniaceae should remain in this order or should be placed in the Peridiniales.

Noctilucales, where the main phase is a large naked cell, often with a tentacle. Reproduction complex.

Pyrocystales, where the main phase is a non-motile coccoid cell and reproduction is by a gonyaulacoid or gymnodinioid motile cell.

Peridiniales, where cells are covered with distinct thecal plates (armour) and in which there is usually a transverse groove (girdle).

Blastodiniales, where the main phase is parasitic and a gymnodinioid motile phase is present.

The classification of the orders, families and genera included in the present work is tabulated below:-

OUTLINE OF CLASSIFICATION AND ARRANGEMENT OF GENERA

Phylum (or Division)	**PYRROPHYTA (or DINOPHYTA)**	
Class	DINOPHYCEAE	Genera
Order	**Prorocentrales**	
Family	Prorocentraceae	Prorocentrum Mesoporos
Order	**Dinophysiales**	
Family	Amphisoleniaceae	Amphisolenia
	Dinophysiaceae	Dinophysis Sinophysis Thecadinium
Order	**Gymnodiniales**	
Family	Gymnodiniaceae	Amphidinium Cochlodinium Gymnodinium Gyrodinium Hemidinium Ptychodiscus Torodinium
	Warnowiaceae	Nematodinium Nematopsides Protopsis Warnowia
	Polykrikaceae	Polykrikos
	Pronoctilucaceae	Oxyrrhis Pronoctiluca
	Lophodiniaceae	Aureodinium Crypthecodinium Endodinium Herdmania Katodinium Sclerodinium Zooxanthella

Genera

Order	**Noctilucales**	
Family	Noctilucaceae	Kofoidinium Noctiluca Spatulodinium

Order	**Pyrocystales**	
Family	Pyrocystaceae	Pyrocystis

Order	**Peridiniales**	
Family	Pyrophacaceae	Helgolandinium Pyrophacus
	Peridiniaceae	Cachonina Coolia Glenodinium Heterocapsa

Diplopsalis 'Group':-

Diplosalis
Diplopsalopsis
Dissodium
Oblea
Zygabikodinium

Scrippsiella
Protoperidinium

	Gonyaulacaceae	Amphidoma Gonyaulax
	Triadiniaceae	Triadinium
	Heterodiniaceae	Heterodinium
	Ceratiaceae	Ceratium
	Oxytoxaceae	Adenoides Amphidiniopsis Corythodinium Oxytoxum Roscoffia
	Cladopyxidaceae	Micracanthodinium Palaeophalacroma
	Podolampaceae	Blepharocysta Podolampas

Order	**Blastodiniales**	
Family	Dissodiniaceae	Dissodinium

KEY TO GENERA

(Mainly based on the morphology of the motile stage)

1a	Cell normally free-living and motile, usually showing some form of transverse groove or cingulum	2
b	Cell free-living (but not motile), parasitic or symbiotic	47
2a	Flagella inserted at anterior pole of cell or near anterior corner. Cell covered with thecal plates: Prorocentrales and Dinophysiales	4
b	Flagella normally inserted on ventral face of cell, girdle usually present, thecal plates may or may not be present	3
3a	Cell 'naked', that is, it lacks substantial thecal plates and readily bursts or deforms upon examination or fixation. If it does not deform then no thecal plates can be seen by light microscopy: Gymnodiniales and Noctilucales	9
b	Cell 'armoured', that is, bounded by a theca consisting of more or less rigid plates, these may be very thin or exceedingly thick and may be extended into projections or horns: Peridiniales	26

Prorocentrales and Dinophysiales

4a	Flagella inserted in small median depression at anterior end of cell; theca mainly consists of two saucer-shaped plates; no girdle	5
b	Flagella inserted at anterior corner (ventral side) of cell; distinct girdle present near anterior end	6
5a	Large pore situated in the centre of each of the two main thecal plates	MESOPOROS
b	Thecal plates smooth, multiple poroid or spiny, anterior spine may be present	PROROCENTRUM
6a	Hypotheca considerably extended, cell usually more than 10 times as long as broad	AMPHISOLENIA
b	Cell not more than twice as long as broad	7
7a	Epitheca reduced in size, only about one third the width of the cell, cell surface lacking ornamentation	SINOPHYSIS
b	Epitheca more or less equal to width of hypotheca, cell surface ornamented with poroids or ridges	8
8a	Girdle lists present, epicone domed	DINOPHYSIS
b	No girdle lists, epicone cap-like	THECADINIUM

Gymnodiniales and Noctilucales

9a	Cell with tentacle or projecting lobe, usually phagotrophic (no chloroplasts present)	10
b	Cell without projections other than flagella	13
10a	Cell very large, over 1 mm diameter	11
b	Cell microscopic	12
11a	Cell with mobile tentacle	NOCTILUCA
b	Cell with rigid tentacle-like projection	SPATULODINIUM
12a	Cell elongated with small anterior tentacle	PRONOCTILUCA
b	Cell helmet-shaped with rounded posterior projection	OXYRRHIS
13a	Organism multinucleate with several sets of flagella	POLYKRIKOS
b	Organism uninucleate, only one set of flagella	14
14a	Cell ∓ symmetrical, girdle median in position	15
b	Cell not symmetrical	17
15a	Cell definitely naked, no shape retained after bursting	GYMNODINIUM
b	Delicate periplast which retains cell shape after examination and fixation	16
16a	Girdle median, sulcus curved to one side of cell, chloroplasts absent	HERDMANIA
b	Girdle median, sulcus straight and slightly incised, chloroplasts present	AUREODINIUM
17a	Girdle short, situated at one side of the cell	HEMIDINIUM
b	No obvious girdle, transparent 'cap' attached to cell	KOFOIDINIUM
c	Girdle complete	18
18a	Girdle situated near one end of cell	19
b	Girdle spirals around cell	20
19a	Girdle near anterior end of cell, epicone reduced	AMPHIDINIUM (ENDODINIUM)
b	Girdle near posterior end of cell, hypocone much reduced	KATODINIUM

20a	Girdle completing only one revolution	21
b	Girdle completing more than one revolution	22
21a	Girdle situated near posterior end of cell, merges with sulcus to form antapical loop	TORODINIUM
b	Girdle ∓ median, theca may be smooth or with longitudinal ridges	GYRODINIUM
c	Girdle ∓ median but incomplete, cell covered with delicate plates, organism found amongst decaying sea-weeds	CRYPTHECODINIUM
22a	Cell without ocellus (conspicuous, large, eyespot)	COCHLODINIUM
b	Cell with ocellus	23
23a	Nematocysts present (large ejectile structures)	24
b	Nematocysts absent	25
24a	Projection from end of girdle; ocellus simple	NEMATOPSIDES
b	No projection from girdle; ocellus lamellate	NEMATODINIUM
25a	Ocellus with several pigment globules	PROTOPSIS
b	Ocellus with distinct pigment cup	WARNOWIA

Peridiniales

26a	Cell epitheca, extended into one long apical horn (or balloon shaped) and one or two antapical horns	CERATIUM
b	Cell spherical to top-shaped, or not as above	27
27a	No defined girdle around cell or, if one present, only has distinct anterior margin	28
b	Clearly defined girdle	30
28a	Cell pear-shaped with apical and antapical protrusions	PODOLAMPAS
b	Cell spherical to ovoid	29
29a	Girdle marked only by an anterior ridge; cell entire	PALAEOPHALACROMA
b	No girdle; two short antapical projections present	BLEPHAROCYSTA
30a	Cell distinctly flattened or distorted	31
b	Cell rounded	35

31a	Cell flattened in anterio/posterior plane; halves of theca become separated	PYROPHACUS
b	Cell flattened in other plane	32
32a	Cell flattened dorso-ventrally	33
b	Cell compressed laterally	34
33a	Cell with median girdle, and may have prominent eye-spot (= KRYPTOPERIDINIUM) GLENODINIUM	
b	Girdle near anterior end, epitheca reduced	AMPHIDINIOPSIS
34a	Cell distorted into elliptical shape, chloroplasts present	COOLIA
b	Cell flattened laterally, epicone reduced to small button	ADENOIDES
c	Cell slightly flattened laterally, epitheca reduced to one quarter width of cell	ROSCOFFIA
35a	Long hair-like projections from surface of cell	MICRACANTHODINIUM
b	Cell without hairs	36
36a	Cell rounded with no projections at apex or antapex	37
b	Apical horns or protruding apical pore, and/or antapical spines present	38
37a	Apical pore structure sigmoid	HELGOLANDINIUM
b	Apical pore structure small, circular	CACHONINA/GLENODINIUM
38a	Cell with single point or dome at apex and antapex	39
b	Cell with two or no projections at antapex	41
39a	Distinct point at apex	OXYTOXUM
b	Apex rounded or domed	40
40a	Girdle slightly displaced, sulcus clearly notches the epitheca	CORYTHODINIUM
b	Girdle median, apex domed, antapex rounded	GLENODINIUM/CACHONINA
c	Girdle near posterior end of cell; wing-like sulcal list present	AMPHIDOMA
41a	Cell spherical to lenticular; distinct curved list associated with sulcus	DIPLOPSALIS 'GROUP'

41b   Cell variously shaped; if sulcus list
        present then antapical spines also present   42

42a   Cell fusiform, small (< 50 μm); thecal
        plates difficult to make out   HETEROCAPSA

b   Cell not fusiform   43

43a   Girdle displaced; only one
        antapical plate   GONYAULAX

b   Girdle straight or only slightly displaced,
        more than one antapical plate   44

44a   Three antapical plates   45

b   Two antapical plates   46

45a   Girdle virtually without lists; thecal
        plates thick but distinction between
        plates difficult to make out   HETERODINIUM

b   Girdle with wide lists; thecal plates clearly
        visible   TRIADINIUM (HETERAULACUS/GONIODOMA)

46a   Thecal plates thin, lacking strong sculpturing;
        girdle plates 6   SCRIPPSIELLA

b   Thecal plates varied, girdle plates 3   PROTOPERIDINIUM

c   Thecal plates varied; girdle plates 5 or 6,
        mainly freshwater   (PERIDINIUM)

Pyrocystales, Blastodiniales etc.

47a   Cells free-living   48

b   Cells symbiotic within animals, chloroplasts
        present, normally non-motile   51

c   Cells parasitic on copepod eggs or animals,
        chloroplasts absent   52

48a   Cell inflated but with distinct transverse groove;
        free-floating in plankton   PTYCHODISCUS

b   Cells of various shapes, no girdle or transverse
        groove visible   49

49a   Cell coccoid; chloroplasts absent; found
        amongst decaying seaweeds   CRYPTHECODINIUM

b   Cells of various shapes, free-floating in plankton   50

50a   Cell crescentic, may contain two to several
        gymnodinioid cells   DISSODINIUM

b   Cell spherical or inflated (entire) or
        crescentic with extended ends   PYROCYSTIS

51a   Non-motile cell spherical with tough wall;
       pyrenoids single-stalked   **ZOOXANTHELLA**

  b   Cell spherical with delicate wall; pyrenoids
       multiple-stalked   **ENDODINIUM**

52a   Parasite on eggs of copepods   **DISSODINIUM**

  b   Ecto-parasite on various planktonic animals
       (not included here: BLASTODINIUM, HAPLOZOON
       OODINIUM)

  c   Endo-parasite within copepods (not included here:
       SYNDINIUM, SOLENODINIUM, etc.)

  d   Parasitic within dinoflagellates (not included here:
       AMOEBOPHYRA)

GLOSSARY

Use in conjunction with Fig. 1

AMPHIESMA. Term used by Schutt (1895) for the thecal covering of dinoflagellates. In this book the term theca will be employed.

ANTAPEX. The posterior end of the cell.

ANTAPICAL PLATES. The plates covering the posterior end of the cell (designated by ' ' ' ').

APEX. The anterior end of the cell.

APICAL PLATES. Those plates which are in contact with the apex or apical pore structure (designated ').

APICAL PORE. A distinctive structure found in several genera, often having a particular shape. It usually consists of a boundary plate and a closing plate and the pore may be elevated on a horn.

ARCHEOPYLE. The aperture in the wall of a cyst through which the contents have emerged. Very frequently it is the position of one or more of the anterior intercalary plates.

AREOLE. An area of a thecal plate demarcated by a ring or ridge.

ARMOURED. Cells with a theca which contains rigid or semi-rigid plates. Often the plates are thickened or ornamented with reticulations etc.

CATENATE. Species which form chains of cells.

CHLOROPLAST. Various shaped bodies which are generally yellow or yellow-brown. Here photosynthesis is carried out enabling chloroplast-containing dinoflagellates to have autotrophic nutrition.

CINGULUM. Another term for the transverse groove or girdle (c.f.).

COCCOID. Spherical cells with what appears to be an entire wall.

CYST. A resistant stage in the life history of a dinoflagellate, probably always formed after fusion of gametes. The cyst wall is often of more than one layer and may be ornamented with spines or other processes.

DISPLACED. Term used to describe a girdle in which the two ends do not meet.

EPICONE. In 'naked' dinoflagellates this is the part of the cell anterior to the girdle.

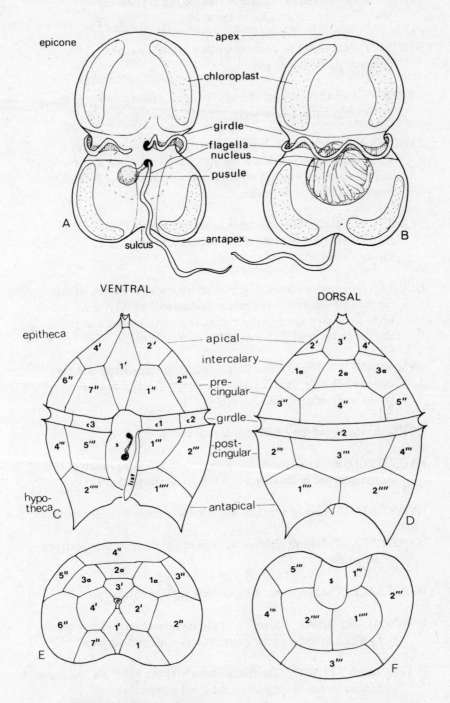

Fig. 1   Diagrams to illustrate glossary:
A. Gymnodinioid cell ventral view.   B. Gymnodinioid cell dorsal view.
C. Peridinioid cell ventral view.   D. Peridinioid cell dorsal view.   E. Peridinioid
cell apical view of epitheca.   F. Peridinioid cell antapical view of hypotheca.
In Figs. C–D the thecal plates are labelled according to the Kofoidean method.

EPITHECA. In 'armoured' dinoflagellates the thecal plates covering the part of the cell anterior to the girdle.

EPITRACT. Name used by palaeontologists for the epicone or epithecal portion of a cyst.

EYESPOT. A small red or orange body found in a fixed position in the cell, usually behind the sulcus.

FLAGELLA. Dinoflagellates have two flagella, a complex helical or sinusoid flagellum which is orientated transversely, and a more normal flagellum which is orientated longitudinally and generally beats in a posterior direction.

FLAGELLAR PORE. The depression in the cell surface ('naked' dino-flagellates) or gap between thecal plates ('armoured' species) through which the flagella emerge.

GIRDLE. A transverse furrow or groove which encircles the cell. It may run straight around the cell or can be displaced by varying degrees up to a complex spiral configuration. It is often termed a *cingulum* and the plates making up the girdle are designated c.

GYMNODINIOID. Naked cells with a median transverse girdle and the usual two flagella. Found as gametes or motile stages in some genera of dinoflagellates.

HORN. Extension of the apical or antapical plates.

HYPNOZYGOTE. A thick-walled zygote formed following fusion of two motile gametes. In dinoflagellates it may also be termed a cyst.

HYPOCONE. The part of a 'naked' dinoflagellate posterior to the girdle.

HYPOTHECA. The thecal plates of an 'armoured' dinoflagellate posterior to the girdle.

HYPOTRACT. The hypocone or hypothecal equivalent of a cyst.

INTERCALARY BAND. The region of the thecal plate adjoining a suture (c.f.) which in some species becomes very wide and obvious.

INTERCALARY PLATES. The thecal plates between either the apical and precingular series or the post-cingular and antapical plates.

LIST. An extension of the thecal plates, usually found around the girdle, where it may be supported by ribs, and along one side of the sulcus where it may protrude as a posterior wing.

LONGITUDINAL FLAGELLUM. See flagella.

LONGITUDINAL GROOVE. See sulcus.

MEGACYTIC. Term used when referring to members of the Prorocentrales
and Dinophysiales which have an enlarged connecting band between
the main thecal plates in preparation for division.

META. Term applied to *Protoperidinium* and related genera and referring
to the five-sided first apical plate.

NAKED. Dinoflagellates with no plates in the theca (or amphiesma).
These cells usually burst very readily and leave no recognisable form.

NEMATOCYST. A large elaborate ejectile organelle reminiscent of those
found in the Coelenterates. Found in the Warnowiaceae and
Polykrikaceae. They appear as ellipsoidal bodies within the cell.

NUCLEUS. The typical dinoflagellate nucleus is large, variously shaped,
and has chromosomes which can easily be seen. Often described as
moniliform or stringy and called a dinokaryon or mesokaryon
because of its unique features.

OCELLUS. A large elaborate version of the eyespot which consists of a
hyaline lens; a dense 'retinoid' and a dark red to black pigment
cup. It presumably functions as a photoreceptor.

ORTHO. Term applied to *Protoperidinium* species with a four-sided
first apical plate.

PARA. Term applied to *Protoperidinium* species with a six-sided
first apical plate.

PARATABULATION. The reflection of thecal tabulation which may often
be seen in the wall of dinoflagellate cysts .

PELLICLE. A tough retaining envelope which may be found around certain
dinoflagellates or which can develop at a certain stage of the life
history.

PLATE. See tabulation.

PROCESS. Spines and other projections, often quite elaborate, found
on dinoflagellate cysts.

PUNCTATE. The thecal plates appear to be covered with pores (punctae).

PUSULE. A unique structure found only in dinoflagellates, usually
associated with flagellar pores. It may appear as a pinkish
sac-like structure. Probably functions in osmo-regulation.

PYRENOID. A dense body associated with the chloroplasts. The core of the pyrenoid is composed of protein and this may be surrounded by starch grains.

RETICULATIONS. (Reticulate). Ridges over the surface of thecal plates, often making a complex pattern.

SCULPTURING. Reticulations, poroids, rings, papillae etc. on the surfaces of thecal plates.

SULCUS. The longitudinal groove or furrow. This may extend onto the epicone or epitheca but more normally runs the length of the hypocone (or hypotheca). In 'armoured' dinoflagellates the sulcal area is covered with a number of small plates.

SUTURE. The junction between adjacent thecal plates.

TABULATION. The arrangement of plates on the surface of 'armoured' dinoflagellates. The basic system, as worked out by Kofoid, consists of a number of series of plates: apicals, precingulars, cingulars, sulcals, postcingular and antapicals; designated ', ", c, s, '", "", respectively. Intercalary plates (c.f.) may also be present.

TENTACLE. A prehensile appendage found in some larger dinoflagellates.

THECA. The complex cell covering of the dinoflagellate cell which consists of several membranes and may include a number of cellulosic plates.

TRANSVERSE FLAGELLUM. See flagella.

TRANSVERSE GROOVE. See girdle.

UNARMOURED. Lacking cellulosic plates in the the theca. See 'naked'.

VALVES. The two main thecal sections covering cells of the Prorocentrales and Dinophysiales. They are orientated longitudinally.

VENTRAL AREA. The central area on the ventral side of *Ceratium* species. It is often shown as lacking any thecal plates but recent studies have revealed the presence of delicate plates.

SELECTED REFERENCE WORKS

General taxonomic works:

Balech, E. (1976). Clave Illustrada de Dinoflagelados Antarticos. Publ. 11. Inst. Antart. Argentino, Buenos Aires. — Keys and descriptions in Spanish and English.

Drebes, G. (1974). Marines Phytoplankton. Thieme, Stuttgart. — Good photographic illustrations of dinoflagellates and diatoms from the Helgoland area of the North Sea.

Kofoid, C.A. & Swezy, O. (1921). The Free-living Unarmored Dinoflagellata. Mem. Univ. Calif. 5, Berkeley. — A very thorough account with useful illustrations.

Lebour, M.V. (1925). The Dinoflagellates of Northern Seas. Marine Biological Association, Plymouth. — The only previous work concerned with the dinoflagellates around the British Isles. Taxonomy rather out of date but nevertheless still useful.

Schiller, J. (1933, 1937). Dinoflagellatae, Vol. 10, pt. 3, I & II. In: Rabenhorsts Kryptogamenflora. Leipzig. — The most comprehensive treatise to date on both marine and freshwater dinoflagellates.

Steidinger, K. & Williams, J. (1970). Memoirs of the Hourglass Cruises: Dinoflagellates. Florida Dept. Nat. Res. Vol. 2. St. Petersburg. — Useful photographs and keys for species found in Florida.

Subramanyan, R. The Dinophyceae of the Indian Seas.
  I. Genus *Ceratium* Schrank. (1968).
  II. Family Peridiniaceae (1971). Marine Biol. Ass. India, Cochin.

Taylor, F.J.R. (1976). Dinoflagellates from the Indian Ocean Expedition. Bibl. Botanica, Vol. 132. Stuttgart. — A useful account of the oceanic species with much discussion of taxonomic problems.

Wood, E.J.F. (1954). Dinoflagellates in the Australian Region. Aust. J. Marine Fresh. Res. Vol. 5, No. 2.

Wood, E.J.F. (1968). Dinoflagellates of the Caribbean Sea and Adjacent Areas. Univ. of Miami, Coral Gables.

Other useful books:

Campbell, P.M. (1973). Studies on Brackish Water Phytoplankton. Sea Grant Publication, Chapel Hill, North Carolina.

Other useful books cont'd

Conrad, W. & Kufferath, H. (1954). Recherches sur les Eaux Saumatres des Environs de Lilloo. II. Partie descriptive, algues et Protistes. Brussels.

Dodge, J.D. (1981). A provisional atlas of the marine dinoflagellates of the British Isles. Natural Environment Research Council.

Lee, A.J. & Ramster, J.W. (Eds.) (1981). Atlas of the seas around the British Isles. M.A.F.F. London.

Lentin, J.K. & Williams, G.L. (1975). A monograph of Fossil Peridinioid Dinoflagellate Cysts. Bedford Inst. Oceanog. Reports.

Lo Cicero, V. (Ed.) (1975). Proceedings of the First International Conference on Toxic Dinoflagellate Blooms. Mass. Sci. Tech. Fndtn.

Sarjeant, W.A.S. (1974). Fossil and Living Dinoflagellates. Academic Press, London & New York. – A good introduction to fossil cysts for biologists.

Sournia, A. (Ed.) (1978). Phytoplankton Manual. Unesco, Paris. – A most useful compilation of methods for collection, examination and enumeration of dinoflagellates and other phytoplankton organisms.

Stover, L.E. & Evitt, W.R. (1978). Analysis of Pre-Pleistocene Organic-walled Dinoflagellates. Stanford Univ. Publ. Geol. Sci., Vol. 15.

Taylor, D.L. & Seliger, H.H. (Ed.) (1979). Toxic Dinoflagellate Blooms. Elsevier/North Holland, New York. The proceedings of the Second International Conference on Toxic Dinoflagellate Blooms.

See p. 273 for other literature cited in the text.

Order PROROCENTRALES

Family Prorocentraceae

MESOPOROS Lillick

Lillick, 1927, p. 497.

Syn: *Porella* Schiller, 1928, p. 54.
       Porotheca Silva, 1960, p. 23.

Cell in valve view round, oval or heart-shaped. Wall consisting of two valves which interlock at their edges, where there may be a rigid thickening, plus a flagellar pore structure inset into a V-shaped depression at the anterior end of one valve (detailed structure unknown). Each valve has a distinct conical depression (the 'pore') situated in a median position, inconspicuous in valve view but very clear in lateral view. This is the character which distinguishes *Mesoporos* from *Prorocentrum.* Flagella, two, inserted at the anterior end of the cell. Cells move forward with a spiralling motion. Chloroplasts of various types, yellow-brown.

Types species: *M. perforatus* (Gran) Lillick.

Note: Six species belonging to this genus have been described but these are all within the same size range and may quite possibly be slight variants of the type species *M. perforatus.* The distinguishing characters, slightly different shape in valve view, more rounded in lateral view; thicker valves; chloroplasts broken up, are all rather poor taxonomic characters for there is always considerable variation of these in any dinoflagellate species, and many of these have been shown to occur in a culture of *M. perforatus* studied by Braarud (1945). We here bring the four species described by Schiller (1933) into synonymy with *M. perforatus* but leave as a distinct species *M. parthasarathicus* Subrahmanyan (1966) which differs from the others by having numerous discoid chloroplasts instead of the two plate-like chloroplasts which they have.

Mesoporos perforatus (Gran) Lillick

Fig. 2A, B, Pl. Ia, b.

Lillick, 1937, p. 497

Syn: *Exuviaella perforata* Gran, 1915.
       Porella perforata Schiller, 1928, p. 55, fig. 12.
       P. asymmetrica Schiller, 1933, p. 27, fig. 29.
       P. adriatica Schiller, 1928, p. 56, fig. 14.
       P. bisimpressa Schiller, 1928, p. 54, fig. 11.
       P. globulus Schiller, 1928, p. 56, fig. 13.
       Porotheca perforata (Gran) Silva, 1960, p. 23.

Having the basic characters of the genus. Cell oval, round or chordate in valve view. Row of small pores around periphery of each valve in addition to central pore. Chloroplasts, two, one situated under each valve, sometimes having a lobed outline. Nucleus situated in the posterior end of the cell.

Size: 14–27 μm long;   18–21 μm wide.

Distribution: Around the British Isles mainly found off the east of Scotland but also off SW England and Wales. Also found off Norway and in the Adriatic Sea. At Plymouth Lebour found it to be abundant in July–October but rare in winter. As a small organism it is easily overlooked and is readily confused with *Prorocentrum* species.

PROROCENTRUM Ehrenberg

Ehrenberg, 1833, p. 307.

Syn: *Cercaria* Michaelis, 1830.
       Exuviaella Cienkowski, 1881.
       Postprorocentrum Gourret, 1883.

Cell rotundate, elliptic, cordiform or lancoelate in valve view, covered mainly by two thecal plates which fit together at their edges; flagellar pore complex, anterior, consisting of a number of small plates which may also bear an anterior spine or projection; flagella two, each emerging from its own pore, one normal and the other helical; nucleus single, in posterior half of cell, containing permanently condensed chromosomes; chloroplasts two, reticulate, one situated under each main thecal plate; pyrenoid(s) sometimes present; sack pusules situated in anterior end of cell, opening into flagellar canal; trichocysts sometimes present, with associated pores through thecal plates; size range from 6–100 μm; probably all marine organisms, world-wide distribution.

Type species: *P. micans* Ehrenberg.

Note: The taxonomy of the genus has recently been revised (Dodge, 1975b) and the number of species considerably reduced. Those included here include only the species which have been found around the British Isles.

KEY TO THE SPECIES OF *PROROCENTRUM*

1a	Thecal plates without obvious pores or any ornamentation	2
b	Thecal plates porous or spiny, anterior projection present or absent	3
2a	Cell ovate in plate view with slight notch at anterior end	*P. aporum*
b	Cell ± elliptic with straight sides and angled anterior end	*P. cassubicum*
3a	Thecal plates apparently covered with minute spines which may appear as poroids under the light microscope (if possible use phase contrast microscopy or iodine-chloralhydrate stain)	4
b	Thecal plates not spiny	7
4a	Posterior end of cell rounded, minute anterior spine(s)	5
b	Posterior end of cell ± pointed, prominent anterior spine or projection	6
5a	Cell spherical or almost so, very small (20 μm)	*P. balticum*
b	Cell ovate to broadly heart-shaped with tiny anterior spine	*P. minimum*
c	Cell ovate, rounded in side view, with two small anterior projections (this species in fact does not have spines but the depressions might be identified as such)	*P. compressum*
6a	Anterior projection a median spine and broad wing; cell broadly heart-shaped	*P. scutellum*
b	Anterior projection pointed or blunt, from one side of the cell; cell lanceolate	*P. dentatum*
7a	Posterior end of cell rounded	8
b	Posterior end of cell pointed or ± acute	9

8a	Anterior end of cell with a distinct depression and *no* spine, cell elongated ovate, widest point behind centre	*P. lima*
b	Anterior end partly depressed but with one or more small spines; cell broadly ovate; surface of plates covered with rows of poroids	*P. compressum*
c	Anterior end of cell not noticeably depressed, cell almost round, without obvious ornamentation, small	*P. nanum*
9a	Cell with anterior projection arising from one side of cell, plates spiny	*P. dentatum*
b	Cell with distinct anterior spine	10
10a	Anterior spine small; thecal plates delicate without obvious ornamentation	*P. triestinum*
b	Anterior spine prominent, long or broad; thecal plates thick and ornamented	11
11a	Spine short with narrow wing	*P. micans*
b	Spine short with broad wing running to margin of cell	*P. scutellum*
c	Spine long, broad and rhombic in side view; cell slender, lanceolate	*P. gracile*

SECTION A

No obvious pores; no anterior spines or ornamentation; no trichocysts

Prorocentrum aporum (Schiller) Dodge

Fig. 2D

Dodge, 1975, p. 107, fig. 1A.

Syn: *Exuviaella apora* Schiller, 1918, p. 258, figs. 12a, b.
        E. grani Gaarder, 1938, p. 73, figs. 9a, b.
        E. antarctica Hada, 1970, p. 11, fig. 3.

Cell broadly ovate, strongly constricted laterally; thecal plates thick and symmetrical, without pores; chloroplasts two.

Size: 30–32 μm long; 21–26 μm wide.

Distribution: Not found during the present survey but previously reported from the North Sea and Norwegian coast. Also found in the Adriatic Sea, Atlantic Ocean, Caribbean Sea, North Sea and Pacific Ocean.

Note: *E. apora* in Lebour (1925) most likely belongs to *P. minimum*. This is a very poorly described species and may well be found to have other characteristics when it can be examined in detail.

Prorocentrum cassubicum (Woloszynska) Dodge

Figs. 2E, F

Dodge, 1975, p. 108, figs. 1B–D, pl. 1A.

Syn: *Exuviaella cassubica* Woloszynska, 1928, p. 251, pl. 3, figs. 6–9.
        Schiller, 1933, p. 22, fig. 18.
        Prorocentrum cassubicum (Woloszynska) Dodge & Bibby, 1973, p. 185, comb. invalid.

Cell ± elliptic with straight sides and the anterior end often angled; thecal plates fairly thick and lacking obvious pores (microscopic pores may be present – Dodge, 1965); two plate-like chloroplasts, supporting large central pyrenoid; trichocysts absent.

Size: 22–25 μm wide.

Distribution: A littoral organism. Not seen during the present survey but Carter (1937) found it in a salt marsh at Brading, Isle of Wight. Also found in the Baltic Sea, and a pool at Woods Hole, U.S.A. from which it has been cultured (see Dodge, 1965).

Note: This has many similarities with *P. lima* but differs slightly in shape and also in the lack of trichocysts.

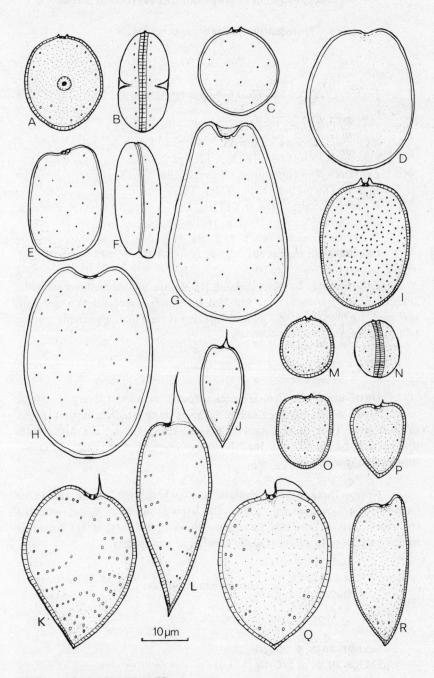

Fig. 2 PROROCENTRALES
A.B. *Mesoporos perforatus.* C. *Prorocentrum nanum.* D. *P. aporum.*
E.F. *P. cassubicum.* G.H. *P. lima.* I. *P. compressum.* J. *P. triestinum.*
K. *P. micans.* L. *P. gracile.* M.N. *P. balticum.* O.P. *P. minimum.*
Q. *P. scutellum.* R. *P. dentatum.*

SECTION B

No anterior spines or ornamentation; trichocyst pores present

Prorocentrum lima (Ehrenberg) Dodge

Figs. 2G, H

Dodge, 1975, p. 109, figs. 1E, F, pl. 1B, C.

Syn: *Cryptomonas lima* Ehrenberg, 1859, p. 793, fig. 73.
       Exuviaella marina Cienkowski, 1881, p. 159, figs. 36, 37.
       Dinopyxis laevis Stein, 1883, p. 11, figs. 27–30.
       Exuviaella lima (Ehrenberg.) Butschli, 1885, p. 151, fig. 2.
       E. laevis (Stein) Schroder, 1900, p. 15.
       E. cincta Schiller, 1918, p. 257, fig. 10.
       E. ostenfeldi Schiller, 1933, p. 18, fig. 12.
       E. caspica Kiselev, 1940, p. 112, figs. 2–3.
       Prorocentrum marinum Dodge & Bibby, 1973, p. 185, comb. invalid.

Cell ± ovate, broadest behind the centre and sometimes with the anterior much narrowed; anterior end indented; thecal valves thick with scattered trichocyst pores; chloroplasts two, plate-like; pyrenoids one or two, large, central; nucleus posterior.

Size: 32–50 μm long; 20–28 μm wide.

Distribution: A littoral species found in almost every sandy or muddy shore which has been sampled, this is undoubtedly a typical species of such sites. It is never found in the plankton. Around the British Isles it appears to be absent from the NE coast of England. Also found in Europe, America, Australasia.

Note: There is some dispute as to whether *E. marina* should be included as a synonym of *P. lima*. The latter is normally shown with the cell sides tapering towards the apex. However, both types are found in similar localities and are most likely varieties of one species.

Prorocentrum nanum Schiller

Fig. 2C

Schiller, 1918, p. 254, fig. 5.
Dodge, 1975, p. 110, fig. 1 H, I.

Syn: *Exuviaella pusilla* (Schiller) Schiller, 1928, p. 50, fig. 4.
       Prorocentrum pusillum (Schiller) Dodge & Bibby, 1973, p. 185.

Cell almost rounded in plate view, broadest at the middle; anterior depression present, slight; thecal plates delicate to thick with scattered trichocyst pores; two chloroplasts, under the plates.

Size: 8–10 μm long 6.5–12 μm wide.

Distribution: The only British record is for a culture established by Dr M. Parke from a water sample collected off Plymouth. Probably a neritic species. Also found in the Adriatic and Atlantic.

SECTION C

Small anterior spines and sometimes surface depressions or ornamentation present; trichocyst pores present

Prorocentrum compressum (Bailey) Abe ex Dodge

Fig. 2I

Dodge, 1975, p. 110, figs. 2F, 4H, I, pl. 4E, F.

Prorocentrum compressum (Bailey) Abe, 1967, p. 373, figs. 2a–d. sine basionym.

Syn: *Pyxidicula compressa* Bailey, 1950, p. 40, pl. 2, figs. 13, 14.
Exuviaella compressa Ostenfeld, 1899, p. 59; 1903, p. 579.
Prorocentrum bidens Schiller, 1928, p. 61, fig. 21.
P. lebourae Schiller, 1928, p. 62, fig. 23.
Exuviaella oblonga Schiller, 1928, p. 50, figs. 6a–c.
E. lenticulata Matzenauer, 1933, p. 438, fig. 1.
E. elongata Rampi, 1951, pl. 1, fig. 9.

Cell ovate to rotundate in valve view, rounded and not much compressed in side view; anterior end usually with very slight depression and two projections, equal, unequal, or bearing fine wings; thecal plates covered with small depressions; chloroplasts two, plate-like.

Size: 30–50 μm long; c. 25 μm wide.

Distribution: Found to the north and north west of the British Isles. As it has also been found in the Atlantic ocean it would suggest that this is an oceanic species which drifts eastwards with the Gulf Stream. Also found in many other parts of the world in oceanic and neritic situations, occasionally embedded in mucilage: Pacific, Atlantic, Mediterranean, North Sea, Australasia, Japan.

Prorocentrum triestinum Schiller

Fig. 2J

Schiller, 1918, p. 252, figs. 1a, b.

Syn: *Prorocentrum redfeldii* Bursa, 1959, p. 19, figs. 121–124.
P. pyrenoideum Bursa, 1959, p. 18, figs. 112–120.

Cell rounded at the anterior end, pointed at the posterior, about twice as long as wide; anterior end with short apical spine and, to one side, a depression in which the flagellar pores are situated; thecal plates with scattered trichocyst pores; nucleus round, posterior.

Size: 18–22 µm long; 6–11 µm wide.

Distribution: This has been recorded at Plymouth, where a culture was established. As a rather small, slim organism it is probably easily missed by most sampling techniques. Also found in the Adriatic, Mediterranean, off the W. coast of Africa and off Japan. Large numbers have been found off the Dutch coast in September (Kat, 1977).

SECTION D

Large anterior spine present; posterior end pointed; surface of plates covered with depressions

Prorocentrum micans Ehrenberg

Fig. 2K, Pl. Ic, d.

Ehrenberg, 1833, p. 307.

Syn: *Cercaria* sp. Michaelis, 1830.
Prorocentrum schilleri Bohm in Schiller, 1933, p. 38, figs. 40a–e.
P. levantinoides Bursa, 1959, figs. 125–127.
P. pacificum Wood, 1963, p. 8, fig. 5.

Cell with a rounded anterior end and pointed posterior end, broadest around the middle or towards the anterior, usually less than twice as long as broad; prominent anterior spine with wings; thecal plates perforated by numerous tubular trichocyst pores, mainly arranged in radial rows, and surface of plates covered by regularly arranged depressions; chloroplasts containing large internal pyrenoids; nucleus large, V-shaped and situated in the posterior end of the cell.

Size: 35–70 µm long, 20–50 µm wide.

Distribution: In the present survey this species has been found in almost every coastal site sampled. It is unusual to find it in samples taken more than about 50 km from the coast. In general it appears to be more abundant in the south and west than in the east or north. It is a cosmopolitan neritic species reported from practically all parts of the world.

Note: This is an extremely variable species and also one which is probably more abundant than any other member of the genus. It would seem that many workers have discovered local forms and raised them to specific rank, thus confusing the literature. It is not easy to distinguish this species from *P. scutellum* on one hand and *P. gracile* on the other.

Prorocentrum gracile Schutt

Fig. 2L

Schutt, 1895, pl. 1, fig. 3

Syn: *P. macrurus* Athanassopoulos, 1931, p. 14, fig. 15.
P. hentschelii Schiller, 1933, p. 37, figs. 38a, b, 'hentscheli'.
P. sigmoides Bohm, 1933, p. 398, fig. 1.
P. diamantinae Wood, 1963, p. 2, fig. 2.

Cell elongate, lanceolate, with rounded anterior end and pointed posterior end; anterior spine long, sharp, narrow in plate view, broad-lanceolate in side view; thecal plates with surface perforated by trichocyst pores and patterned with surface depressions.

Size: 40–60 μm long.

Distribution: Found in a net sample taken off St. Andrews, Scotland in 1971. Also found in the Atlantic, Pacific, Mediterranean.

Note: This species has certain similarities to *P. micans* and Steidinger & Williams (1970) show photographs of the thecal plates in which the pattern of trichocyst pores appears identical with that of *P. micans.* It is distinguished by having a much longer anterior spine and by being at least twice as long as broad.

Prorocentrum scutellum Schroder

Fig. 2Q

Schroder, 1900, pl. 1, fig. 12.

Syn: *P. sphaeroideum* Schiller, 1928, p. 61, fig. 25.
P. robustum Tafall, 1942, pl. 34, figs. 9, 10.

Cell broadly heart-shaped with rounded or pointed posterior end; anterior end with slight indentation and two spines, to the larger of which is attached a delicate broad wing; valves with large trichocyst pores and small poroids, together with small depressions, scattered all over the surface (Lebour); thecal plates thick; chloroplasts two; nucleus U-shaped, at posterior end of cell.

Size: 35–57 µm long;   30–45 µm wide.

Distribution: Found off the South West coast of Scotland, also in the Adriatic, Atlantic (E. and W.), Arctic Sea, Indian Ocean, Australasia.

Note: Superficially this organism bears a strong resemblance to the short forms of *P. micans*, and the nucleus, as shown by Schiller (1928), and trichocyst pores, as shown by Bursa (1959), appear identical with those of *P. micans.* However, Lebour (1925) and Schiller (1933) both refer to the presence of small spines over the surface of the plates, and if this is true it would clearly separate the organism from *P. micans.* But the supposed spines are almost certainly depressions like those found in *P. micans* and *P. compressum.* Like *P. gracile* it appears to be an intermediate form. The more easily recognised anterior wing is the main character that has traditionally been used to characterise this species.

SECTION E

Spiny thecal plates present; trichocyst pores and anterior spine may also occur

Prorocentrum dentatum Stein

Fig. 2R

Stein, 1883, pl. 1, figs. 14, 15.

Syn: *Prorocentrum obtusidens* Schiller, 1928, p. 57, fig. 15.
       P. veloi Tafall, 1942, p. 437, figs. 4–6.
       P. monacense Kufferath, 1957, figs. 1, 2.

Cell elongated, heart-shaped to lanceolate with an anterior extension at one side which may be pointed or blunt; anterior end sometimes with small depression; posterior end acute; plates with surface covered with small spines.

Size: 36–60 µm long; 15–20 µm wide.

Distribution: Thought to be an oceanic species. The only records from around the British Isles are from a culture established at Plymouth by Dr. M. Parke (see Dodge 1965 for electron microscopy of its thecal plates), and a cell observed in a sample from Galway Bay, Ireland. Also found in the Atlantic, Pacific, North Sea, Mediterranean, Sargasso Sea and off Japan.

Prorocentrum minimum (Pavillard) Schiller

Fig. 20P, Pl. I f, g.

Schiller, 1933, p. 32, figs. 33a, b.

Syn: *Exuviaella minima* Pavillard, 1916, p. 11, p.1, figs. 1a, b.
Prorocentrum triangulatum Martin, 1929, p. 557, figs. 1–3.
Exuviaella mariae-lebouriae Parke & Ballantine, 1957, p. 645, figs. 1–9.
Prorocentrum cordiformis Bursa, 1959, p. 31, figs. 104–107.

Cell ovate, triangular or heart-shaped in plate view, posterior end usually rounded and anterior end truncate with a very slight depression, anterior spine small (not always visible in the light microscope); trichocyst pores mainly around margin of plates, surface of plates covered with minute spines; chloroplasts with associated pyrenoids.

Size: 14–22 μm long; 10–15 μm wide.

Distribution: Our only records are from the culture (named *Exuviaella mariae-lebouriae*) established at Plymouth (Parke & Ballantine, 1957); from a collection made from the shore at Littlehampton, Sussex; and another, of the *triangulatum* form, from near Bognor, Sussex. Also found commonly among the west coast of U.S.A., in Japan, Gulf of Mexico, Caspian Sea and the Mediterranean, often in very large numbers.

Note: As a small species this is probably frequently lost or over-looked in sampling.

Prorocentrum balticum (Lohmann) Loeblich

Figs. 2M, N, Pl. I e.

Loeblich, 1970, p. 906.

Syn: *Exuviaella baltica* Lohmann, 1908, p. 265, pl. 17, figs. 1a, b.
Prorocentrum pomoideum Bursa, 1959, figs. 108–111.
Exuviaella aequatorialis Hasle, 1960, p. 29, figs. 18a, b.

Cell round to slightly ovate in valve view, round and scarcely com-pressed in side view, with minute apical projections beside flagellar pores; thecal plates covered with minute spines; nucleus sub-spherical and posterior in position.

Size: 9–10 μm long; 7–20 μm wide.

Distribution: This species was recorded in several parts of the North Sea by Grontved (1952) but we have only found it off western Scotland in the Gulf Stream area. It has been isolated into culture at Plymouth.

Note: This very small *Prorocentrum* is sometimes present in very large quantities. Blooms have been reported from the Baltic Sea, Norwegian fjords, the Atlantic off Angola, the East Coast of the U.S.A., and the Japanese coast. It has been seen in numerous other places in spite of the fact that its small size means that it is not usually collected in net samples.

Order DINOPHYSIALES

Family Amphisoleniaceae

AMPHISOLENIA Stein

Stein, 1883, p. 9, 24.

A genus of very elongated organisms, epitheca reduced to a small cap-like plate, hypotheca drawn out into one or more posterior processes which may be inflated in their upper or mid-portion. Transverse lists as in *Dinophysis* but sulcal list reduced and only running a short distance along the hypotheca. Thecal plates weakly sculptured. Essentially a tropical genus but occasionally found in the north in Gulf Stream waters.

Type species: *A. globifera* Stein.

Amphisolenia globifera Stein

Fig. 3E

Stein, 1883, p. 24, pl. 21, figs. 9—10.
Schiller, 1933, p. 174, fig. 161.

Epitheca very reduced; hypotheca of variable length, widened into a bulge around the centre and ending with a rounded knob. Sulcus less than a quarter of the length of the hypotheca.

Size: 200–300 µm long.

Distribution: Found in N. Bay of Biscay (Gaarder, 1954), warm part of Atlantic, Mediterranean.

Family Dinophysiaceae

DINOPHYSIS

The genus *Dinophysis* was introduced by Ehrenberg in 1840 with *D. acuta* as the type species. This was the only genus in the family until 1883 when Stein proposed *Phalacroma* to include those species with a clearly visible prominent epitheca, *P. porodictyum* being the type species. In 1923 Jorgensen placed *P. nasutum* in a new genus *Pseudophalacroma*. He then divided *Dinophysis* and *Phalacroma* into sections based on different morphological characters, using Pavillard's sections where appropriate. The separation into the two genera was not always easy and these difficulties were again encountered by Kofoid & Skogsberg (1928). The classification put forward by Kofoid & Skogsberg incorporated many of the sections of Jorgensen and was adopted by Schiller in Rabenhorst's Kryptogamenflora (1933). Tai & Skogsberg (1934) were the first to describe the plates, including those of the sulcus and noted that *Phalacroma* and *Dinophysis* "are very closely related, and their generic separation is questionable". They recorded three plate features shared by both genera:

1.   All four sulcal plates border the flagellar pore.

2.   Both the ventral hypothecal plates are small and elongate, arranged one behind the other, with their long axes in the longitudinal plane.

3.   The two ventral epithecal plates are similar in shape, size and position.

Abe (1967a, b) confirmed this plate arrangement and established the relationship between the left sulcal list and the ventral hypothecal plates. Balech (1944) introduced *Prodinophysis* as a synonym for *Phalacroma* but this term was not widely accepted although Loeblich III transferred several *Phalacroma* spp. into the new genus in 1965. Abe (1967b) in describing some *Phalacroma* species included them in the genus *Dinophysis*. Since this time the single genus *Dinophysis* has been used to include those species previously classified under *Phalacroma*.

Description of the Genus

Dinoflagellates which are frequently laterally flattened, although some nearly spherical species have been described viz. *D. contracta*. Usually divided into a small epitheca, a moderately wide cingulum or girdle and a hypotheca which frequently represents three-quarters or more of the total cell length. The epitheca may be flat and low or may be rounded to form a slight dome or a prominent one which rarely equals the hypotheca in length and never exceeds it. The cingulum is not displaced and may be excavated but is always bordered by lists which may be equal in width or the anterior one may be larger than the posterior one, may be supported by ribs, and may be horizontal or projected anteriorly. The hypotheca may be ornamented with spines or lists (sometimes referred to as sails).

The ventral sulcus which is found on the hypotheca but which can extend onto the epitheca is composed of four plates surrounding the flagella pore. It usually extends no more than halfway down the length of the hypotheca and is bordered by lists formed from hypothecal plates. The left sulcal list made up of lists from H_1 (the anterior part) and H_2 (the posterior part) is supported by three ribs (although no ribs are found in *D. contracta*), the anterior main rib (R_1), the fission rib (R_2) and the posterior main rib (R_3). The right sulcal list is not supported by ribs and is the list on H_3. These lists may be very small or may be very elaborate.

The theca may be porate and have areoles or reticulations. The nucleus is large and situated in the region of the hypotheca; a pusule may be present and chromatophores may also be present. Megacytic stages frequently occur: for a description of the megacytic zone see Abe (1967a, 382–387) and Tai & Skogsberg (1934, 401–404). Division, when it occurs, is by binary fission and two daughter cells still connected are not uncommon in plankton samples. No account of sexual reproduction in this family is known. No toxicity or bioluminescence is associated with members of this genus. Much has yet to be learned of this genus but difficulties in culturing are holding back progress in this direction. However, some interesting work has been carried out by Solum and others on effects of environment on cell size. Working with *D. lachmanii* (= *D. acuminata*) Solum (1962) found that in areas of low salinity cells were longer than in areas of higher salinity. A similar effect of salinity was recorded by Matzenauer (1933) working on cells from the *'caudata'* group. Solum also looked at temperature and found that low temperatures led to relatively longer cells but where low salinity and high temperatures existed the salinity effect predominated.

Problems in identification arise from inadequate descriptions of species, poor drawings and the most frustrating variations within a clone. In some instances, viz. *D. ovum* and *D. lenticula*, it is virtually impossible to delimit the species morphologically. Some numerical taxonomic work on relatively large numbers of individuals in a population has been published by Solum (1962). Although this illustrates the norm it also shows the wide variation that can occur within a species.

Thecal Plates (Fig. 3A–C)

When first described, members of this genus were thought to consist of two valves with pairs of epithecal, cingular, and hypothecal plates arranged bilaterally. Tai & Skogsberg (1934) were the first to describe the thecal plate arrangement in the Dinophysoidae and to explain the relationships between the plates. Since then further studies have been made on the plates notably by Abe (1967a, 1967b), Balech (1957, 1976a, 1976b) and Norris & Berner (1970).

Seventeen plates were originally thought to be present, then Norris & Berner (1970) described a further epithecal plate in their paper on the Hastata group of Dinophysis. The epitheca is composed of six plates, and the cingulum, hypotheca and the sulcus four each.

The epithecal plates are left and right dorsal epithecal plates (E_2 and E_3), the left and right ventral epithecal plates (E_1 and E_4), the pore plate (Po) and a pore platelet (P_{p1}). E_2 and E_3 make up most of the area covered by the epitheca and are connected by a sagittal fission suture. They have a list on their lateral edge which forms the anterior cingular list. E_1 and E_4 are much smaller than the two dorsal epithecal plates and each has a list which connects with the list of the corresponding dorsal epithecal plate. Ventrally they touch two sulcal plates. The pore plate (Po) is smaller than the ventral epithecal plates and is situated in a notch in E_2 and E_3. Normally it lies dorsal to the apical pore but in *D. odiosa* a pore is present in Po. The pore platelet is smaller than Po; indeed so small that it is frequently overlooked, and lies to the left and/or ventral to Po. The cingular plates are left and right dorsal cingular plates (C_2 and C_3) and left and right ventral cingular plates (C_1 and C_4). The dorsal cingular plates are larger than the ventral cingular plates. C_1 can be larger than C_4. The hypothecal plates are left and right dorsal hypothecal (H_2 and H_3) and left and right ventral hypothecal plates (H_1 and H_4). H_2 and H_3 are the largest plates of the theca and may determine the shape of the body. They are joined at the sagittal fission suture from the cingulum on the dorsal side to the ventral notch of H_3. Each dorsal hypothecal plate has a list along its anterior edge forming the posterior cingular list. Ventrally this list on H_2 joins the list on H_1; the list on H_3 continues to the posterior as the right sulcal list which may be short or may terminate at the posterior end of the sulcus. H_3 also has a list behind the posterior main rib (R_3) and is a continuation of the left sulcal list. Members of the "hastata" group also have a posterior list or sail on H_3. H_1 amd H_4 are narrow elongated plates. The right lateral margin of H_1 connects with the anterior portion of the posterior and left sulcal plates. H_4 rests in a notch in H_3 and lies almost immediately behind H_1 so that its left margin is in contact with H_2 at the sagittal fission suture. The right margin is attached to Sp and H_3. The anterior portion of the left sulcal list is composed of a list belonging to H_1 and the posterior portion of a list from H_4.

The sulcus is composed of the anterior and posterior sulcal plates (Sa, Sp) and the left and right sulcal plates (Ss, Sd). Sa is narrow, elongate and touches E_1 and E_4 anteriorly, laterally C_1 and Ss and Sd, and borders the flagellar pore posteriorly. Ss, the smallest sulcal plate, touches C_1 anteriorly, H_1 and Sa and the flagellar pore laterally and Sp and the flagellar pore posteriorly. Sd runs along the right side of Sa touching E_4 anteriorly C_4, H_3 and Sa laterally and surrounds the left anterior part of the flagellar pore posteriorly. Sp is the largest sulcal plate and forms the posterior part of the flagellar pore so that it is concave anteriorly. Laterally it touches H_3, H_1 and H_4, and posteriorly H_3.

KEY TO THE SPECIES OF *DINOPHYSIS*

1a	Cell with posterior sail, with or without spine	2
b	Cell with pointed or rounded hypotheca with or without small 'knobbly' protuberances	3
c	Cell with hypotheca drawn out into large single or double tapering antapical process; wide left sulcal list	*D. caudata*
2a	Epitheca clearly visible, not masked by anterior cingular list. Posterior sail relatively small	*D. mucronata*
b	Epitheca masked by funnel-like anterior cingular list. Posterior sail large and may be supported by a rib	*D. hastata*
3a	Cell with left sulcal list not supported by ribs. Small species with clearly-visible epitheca	4
b	Cell with left sulcal list supported by three ribs	5

4a   Epitheca notched in lateral view; sulcus
      extending onto epitheca   *D. nasutum*

 b   Epitheca rounded and smooth; sulcus not
      extending onto epitheca   *D. contracta*

5a   Cell with hypothecal point ventral to the
      midline. Greatest width of cell below centre of
      hypotheca   *D. acuta*

 b   Cell pointed at midline or rounded symmetrically
      when viewed laterally   6

6a   Cell with rounded epitheca visible beyond
      anterior cingular list   7

 b   Cell with epitheca obscured by anterior cingular
      list   8

7a   Body spherical, covered with poroids, left
      sulcal list narrow. Relatively broad cingulum,
      deeply excavated, <36 μm long   *D. pulchella*

 b   Laterally flattened, with medium sized left
      sulcal list, > 36 μm long   *D. rotundata*

8a   Hypotheca ending in a point; not rounded   9

 b   Hypotheca rounded or may have small antapical
      protuberances   10

9a   Cell with ± parallel sides to hypotheca. Long
      left sulcal list > ¾ length of hypotheca   *D. dens*

 b   Left sulcal list <¾ length of hypotheca.
      Cell widest at middle. Ventral margin of main
      (dorsal) hypothecal plates concave. Point of
      hypotheca in midline   *D. norvegica*

10a   Cell much longer than broad with or without
      posterior protuberances   *D. acuminata*

 b   Cell ovoid and rounded posteriorly   11

11a   Left sulcal list not deflected to right.
      Greatest width at centre of hypotheca   *D. ovum*

 b   Left sulcal list deflected to right   12

12a   Cell small (<40 μm), regularly ovate   *D. punctata*

 b   Cell larger (> 40 μm) broadest near the posterior
      end   *D. recurva*

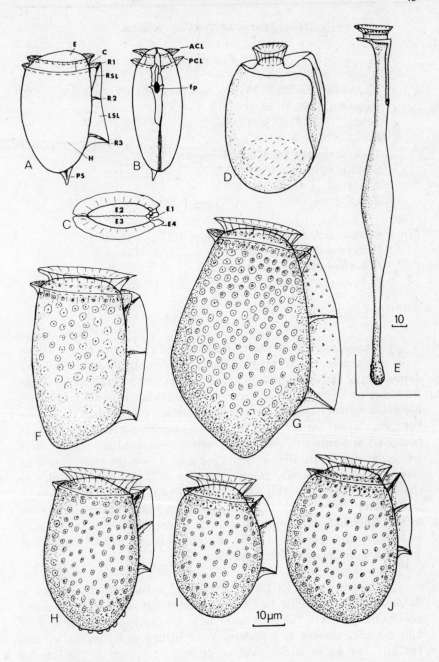

Fig. 3 DINOPHYSIALES

A.)   – Left side view
B.) diagrams to show plates   – Ventral view
C.)   – Anterior view

D. *Sinophysis ebriolum*. E. *Amphisolenia globifera*. F. *Dinophysis dens*.
G. *D. acuta*. H. *D. acuminata*. I. *D. punctata*. J. *D. ovum*.

All originals except A–C after Norris and Berner 1970.

Key to lettering: E = Epitheca. H = hypotheca. C = cingulum (girdle).
R_1 = anterior rib. R_2 = middle rib. R_3 = posterior rib. RSL = right sulcal
list. LSL = left sulcal list. PS = posterior spine. fp = flagella pore.
ACL = anterior cingular list. PCL = posterior cingular list.

Dinophysis acuminata Clap. & Lach.

Fig. 3H

Claparede & Lachman, 1959, p. 408, pl. 20, fig. 17.
Jorgensen, 1899, p. 30, pl. 1. figs. 7–9.
Lebour, 1925, p. 80, pl. 12, figs. 2a–c.
Kofoid & Skogsberg, 1928, p. 59, 219, 224, 228, fig. 2.
Tai & Skogsberg, 1934, p. 433, figs. 4A–T.
Schiller, 1933, p. 120, figs. 113a–g.
Abe, 1967, p. 43, figs. 7a–g.
Balech, 1976, p. 85, figs. 3A–O and T–V.

Syn: *Dinophysis borealis* Paulsen, 1949, p. 46, figs. 14K–U and 15.
      D. lachmanii Paulsen, 1949, p. 46, figs. 14A–H and 15.
      D. boehmi Paulsen, 1949, p. 45.
      D. lachmanii Solum, 1962, p. 9, figs. 2, 1–16, figs. 5, 4–6,
                                              figs. 9, 1–16.
      D. skagii Paulsen, 1949, p. 48, figs. 14, 15.

A variable species, in lateral view elongated and narrow to oval with or without ventrally placed antapical protuberances. Epitheca low, reduced dorsoventrally and either flat or weakly convex but generally sloping from the higher dorsal end to the ventral end. Masked in lateral view by the higher, funnel-like anterior cingular list. Cingulum deeply excavated on dorsal side. Posterior cingular list is less wide than the anterior one and inclined to be horizontal in contrast to the strongly anteriorly-inclined anterior list. Both are supported by small ribs. In lateral and ventral view the main hypothecal plates are generally weakly convex. However, in lateral view they may be almost straight-sided, or with a straight dorsal edge and convex ventral edge or vice versa. The antapical region is rounded and either smooth or with one to four knob-like protuberances of irregular size. The left sulcal list is of medium width, supported by three ribs and extending to just over halfway down the ventral margin of the hypotheca.

Between R_2 and R_3 the list is deflected to the right. R_2 is closer to R_1 than to R_3; the first two being deflected anteriorly and R_3 curved towards the posterior. The right sulcal list is reduced and triangular in shape with the apex situated at a position corresponding to R_2 on the left list. The lists may be smooth or sculpturing may be present in the form of irregular linear markings or areoles. The epithecal and hypothecal plates are covered with areoles which may be coarse or weakly developed, densely packed together or more spatially arranged and each may contain a pore. It is more usual for each areole or poroid to contain a pore but scattered pores with approximately one in three poroids with a pore may occur.

Size: 38–58 µm long; length/breadth ratio 1:3; range 1.29–1.6.

Distribution: Occurs all round Britain except off extreme SE coast. Mostly in spring and summer. Elsewhere recorded widely.

Note: *Dinophysis skagii* Paulsen (1949) reported also by Solum (1962) and Balech (1976a) may be an aberrant form of *D. acuminata* or a separate species. To date few examples have been seen. It is similar to *D. acuminata* but differs in having a ventral outline more sharply curved than the dorsal when seen in lateral view. There are two antapical knobs, it is 35–38 μm in length. See Balech (1976a) for sulcal plate description.

Dinophysis acuta Ehrenberg

Fig. 3G, Pl, II e

Ehrenberg, 1840, p. 18, pl. 1–4.
Ehrenberg, 1839, p. 124, 151, pl. 4.
Lebour, 1925, p. 79, pl. 12, fig. 1.
Schiller, 1933, p. 131, figs. 124a–j.
Solum, 1962, p. 31, fig. 19.
Balech, 1976, p. 80, figs. 1N–W, 2A–D.

Cell flattened laterally, in ventral view sides slightly convex; ovoid in lateral view with the posterior forming a blunt projection directed ventrally. Greatest width of cell in posterior half of hypotheca. Anterior two-thirds of hypotheca with gently convex sides, posterior third with straight dorsal edge and slightly concave ventral edge. Cingular lists are well developed, the anterior one obscuring the low epitheca which may be flat or gently convex. The left sulcal list is broad, approximately one-third of the main hypothecal plate in width.

The main epithecal and hypothecal plates are strongly areolated, each areole containing a pore. The left sulcal list is slightly areolated. There are three sulcal ribs, R_1 pointing towards the anterior end, R_2 ventrally and R_3 to the posterior end. Chromatophores are yellow.

Size: 54–94 μm long; greatest depth, i.e. dorsoventral length 43–60 μm.

Distribution: Found throughout the year but more freqently in spring and summer around British Isles but significantly not reported from the English Channel. Worldwide distribution.

Dinophysis caudata Saville-Kent

Fig. 4B, Pl, II d

Saville-Kent, 1881, p. 455, 460.
Jorgensen, 1923, p. 24, figs. 30–34.
Lebour, 1925, p. 82, fig. 21c.
Kofoid & Skogsberg, 1928, p. 314, figs. 44–45.
Schiller, 1933, p. 153, figs. 145a–o.
Tai & Skogsberg, 1934, p. 453, figs. 9A–K, 10D–F.
Abe, 1968, p. 56, figs. 14a–d.

Syn: *Dinophysis caudata* var. *tripos* (Gourret) Gail, 1950, p. 27, pl. 5, fig. 7.
    D. tripos Gourret, 1883, pl. 3, fig. 53.
        Jorgensen, 1923, p. 29, figs. 38–39.
        Lebour, 1925, p. 82, fig. 22.
        Schiller, 1933, p. 159, figs. 146a–g.
        Tai & Skogsberg, 1928, p. 456, figs. 10A–C.
        Balech, 1944, p. 436, pl. 4, figs. 42–45, pl. 5, figs. 46–45.
    D. homunculus Stein, 1883, p. 3, 24, pl. 21, figs. 1–7.

Distinctive species with long ventral hypothecal projection. In forms previously described as *D. tripos* with additional shorter, dorsal projection. Much variation in size and shape of this dorsal prolongation of the hypotheca has been observed by various workers, resulting in differnet subspecies, varieties or forms being created. Since this species occurs widely more work is required to determine which factors affect the morphology. The epitheca is obscured by a deep funnel formed by the anterior cingular list and supported by ribs. At the distal margin the funnel flattens out. The posterior cingular list is narrow dorsally, becoming wider on the left ventral side where it meets the left sulcal list. The expansion in width is less marked on the right ventral side where it joins the right sulcal list. The posterior cingular list is projected anteriorly and is supported by ribs. Viewed laterally, the hypotheca is narrow at the girdle; the ventral margin is usally straight or sigmoid to the base of the left sulcal list from where it bends sharply towards the centre before again sharply bending towards the posterior; the dorsal margin is straight or concave near the girdle, becoming straight or convex towards the posterior. The hypotheca has its widest point usually at the position of the base of the left sulcal list. The dorsal side of the hypotheca may curve sharply towards the centre where it turns to continue down the drawn-out posterior part of the hypotheca or may itself be projected to form a hollow point. The transition between the anterior and posterior portions of the hypotheca may be sudden or gradual. The epithecal and hypothecal plates are covered with areoles each containing a pore or, at the plate margins, one or two pores. The left sulcal list is wide and supported by three ribs spaced equally apart, which are straight or slightly curved. R_1 projects anteriorly, R_2 ventrally and R_3 posteriorly. The right sulcal list is widest at its anterior end, becoming very narrow between the position R_2 and R_3 where it ends. The lists may be reticulated. Twinned forms occur frequently, joined together on the dorsal surface at the widest part of the hypotheca.

Size: 70–170 μm long.

Distribution: Found off SW England and North of Scotland in late spring and summer.

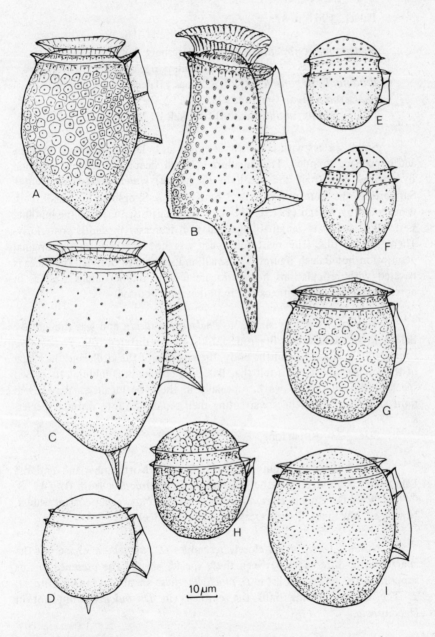

Fig. 4   **DINOPHYSIS**
A. *D. norvegica*. B. *D. caudata*. C. *D. hastata*. D. *D. mucronata*.
E.F. *D. pulchella*. G. *D. nasutum*. H. *D. contractum*. I. *D. rotundatum*.

Dinophysis contracta Kofoid & Skogs.

Fig. 4H, Pl. II b.

Balech, 1967, p. 82.

Syn:	*Phalacroma contractum* Kofoid & Skogsberg, 1928, p. 83, fig. 3, 1.
							Schiller, 1933, p. 62, fig. 55.
		Prodinophysis contracta (Kof. & Skogs.) Balech, 1944, p. 429.
		Phalacroma ruudii Braarud, 1935, p. 112, fig. 32.
		Prodinophysis ruudii (Braarud) Loeblich III, 1965, p. 17.

Small species with conspicuous epitheca. In lateral view both epitheca and hypotheca rounded. The cingulum is broad, deeply incised and bordered by narrow lists (half the width of the girdle) which are sub-horizontal. Supporting ribs were not described by Kofoid & Skogsberg but small teeth were noted by Balech (1973). The cell is deepest just anterior to the midline. Left sulcal list 0.44 length of the cell and decreases in width posteriorly. There are no ribs. The right sulcal list was not described in the original description but Balech figures it as small and sigmoid in outline. The theca is covered with reticulations but pores are few in number and scattered. In dorsoventral view the epitheca and hypotheca are pointed.

The species is very similar to *Phalacroma ruudii* and was thought by Balech to be conspecific. Braarud said his species differed from *Phalacroma contractum* in 'the shape of the body, the position of the girdle and the shape of the left sulcal list'. It is felt that Balech is correct in combining these two species although further work, especially in the type localities, may reveal more significant differences warranting their separation into distinct species.

Size:	23–30 μm long.

Distribution:	From the North Sea off Northumberland and off Lancashire. Kofoid & Skogsberg described *P. contractum* from 07° 47' S, 94° 5' W in December. *P. ruudii* recorded from Norwegian coastal waters, North Sea, NW Spain, Pacific Ocean and North of Iceland.

Note:	*D. contracta* closely resembles *D. pulchella* in shape but the characteristic differences between these species are: 1. The presence of ribs supporting the left sulcal list in *D. pulchella;* these are absent in *D. contracta.* 2. The sulcus extends onto the epitheca in *D. pulchella* but not in *D. contracta.*

Dinophysis dens Pavillard

Fig. 3F

Pavillard, 1915, p. 122, 1916, p. 57, pl. 3. fig. 1.
Jorgensen, 1923, p. 21, fig. 23.
Schiller, 1933, p. 130, fig. 123.

Paulsen, 1949, p. 50, figs. 16a–c.
Solum, 1962, p. 32.
Balech, 1976, p. 90, figs. 4A–I.

Cell with long almost straight-sided main hypothecal plates in lateral view with long left sulcal list and obscured epitheca. Epitheca flat or slightly rounded. Girdle not markedly constricted although the dorsal margin may be gently concave. Girdle lists well developed, inclined anteriorly, the anterior one being at least twice as deep as the posterior. No supporting ribs recorded. Maximum depth of hypotheca near middle. Dorsal and ventral margins of main hypothecal plates straight-sided or gently convex for anterior two-thirds. In posterior third the dorsal margin keeps either the same convexity or bends slightly towards the ventral side, whereas the ventral margin bends sharply towards the dorsal side. The two sides meet in a blunt point a little to the dorsal side of the midline. In ventral view the hypotheca is shaped as a narrow convex wedge.

The left sulcal list is long and extends to the point where the ventral margin has its sharp dorsally-directed bend. Three supporting ribs are evenly spaced. R_1 projects anteriorly, R_3 posteriorly (either curved or straight) and R_2 may be directed either way. The anterior part of this list is heavily sculptured. The right sulcal list ends at about the position of R_2. Surface of theca covered with poroids. Balech (1976) who gives a detailed description of all the plates did not record pores in the poroids.

Size: 50–59 µm long; maximum width of left sulcal list 6–10 µm.

Distribution: Rare to the west of the British Isles; also recorded from the Mediterranean, Danish Strait and off the coast of Norway.

Dinophysis hastata Stein

Fig. 4C, Pl. II g.

Stein, 1883, pl. 19, fig. 12.
Lebour, 1925, p. 83, fig. 12e.
Schiller, 1933, p. 183, fig. 131.

Subovate in lateral view with characteristic elongated triangular sail at the posterior end of hypotheca, ventral of the midline. The epitheca, which is concealed in lateral view by the anterior cingular list, is flat and highest dorsally, sloping to the ventral side. The anterior cingular list is projected anteriorly and supported by numerous ribs; it is relatively wide, at least as wide as the cingulum, raised ventrally but curves towards the horizontal dorsally. The posterior cingular list is narrower. The cingulum is not excavated. The hypotheca is deepest around the middle and the posterior end is narrower than the anterior end. The posterior sail situated just ventral of the midline is moderately large, usually around 16 µm long and is deflected ventrally. In some cells there is a central supporting rib. The left sulcal list is conspicuous with its maximum width, around 14 μm, at R_3. The ribs are strong, the front two directed anteriorly and the posterior one (R_3) either straight or curved posteriorly. R_2 is nearer R_1 than R_3. The list is covered with an irregular reticulum, and extends for over two-thirds the length of the hypotheca. The right sulcal list is less obtrusive and ends to the anterior of R_3. The hypotheca is covered with areoli and scattered pores. In ventral view the cell is broadly lenticular.

Size: 63–70 μm long; 32–48 μm wide.

Distribution: In Gulf Stream waters, off the North of Scotland and NE and SW England in summer. Elsewhere recorded from Norwegian waters, off NW Spain, and in the Atlantic.

Note: There is confusion between this species and *D. odiosa* (Pavillard) Tai & Skogsberg. Kofoid & Skogsberg (1928) regarded *D. hastata* as a collective species and illustrated many diverse forms which they thought subsequent workers might divide into two or more systematic units. One of these forms was subsequently established as *Phalacroma odiosum* by Pavillard (1931). This species was then put in the genus *Dinophysis* by Tai & Skogsberg (1934). Meanwhile, Schiller, in his monograph of 1933, had reduced *P. odiosum* to a synonym of *D. hastata*. Norris & Berner (1970) gave drawings of all the plates of both *D. hastata* and *D. odiosa*, and Balech (1976a) elaborated on the plate description of *D. odiosa*. Both thought that they were two distinct species. It is felt that the differences, apart from those in the cingular and sulcal plates, are not immediately discernible, especially since the *D. hastata* of Noris & Berner does not resemble that of the type species. The specimens recorded from the waters around Britain are included in *D. hastata* although some are in the size range of *D. odiosum*.

Dinophysis mucronata(Kof. & Skogs.) Sournia

Fig. 4D

Sournia, 1973, p. 22.

Syn: *Phalacroma mucronatum* Kofoid & Skogsberg, 1928, p. 172, fig. 22:
4, 6, 8. Schiller, 1933, p. 98, figs. 89a–c
Phalacroma paulseni Schiller, 1928, p. 67, fig. 29.
Dinophysis paulseni (Schiller) Balech, 1967, p. 88, pl. 1, figs. 13–19.
Norris & Berner, 1970, p. 172, figs. 67–78.

Small species subrotund in lateral view and deepest at the middle with an antapical list. The epitheca is rounded and about 0.1 of the cell long and clearly visible above the anterior cingular list. The cingular lists are subhorizontal or inclined anteriorly. The anterior list is slightly wider than the posterior one and both are smooth. The cingulum itself is not excavated and is approximately 4 μm wide. The hypotheca has a rounded outline and in ventral view the body is subovate and symmetrical. The left sulcal list extends just over halfway down the hypotheca and is widest at R_3. R_2 is closer to R_1 than R_3. R_1 and R_2 are straight or slightly concave and directed anteriorly; R_3 is straight and inclined posteriorly. R_3 is not readily discernible but Norris and Berner (1970) record that it ends just in front of R_3. Both sulcal lists are smooth. The antapical list is triangular and follows the longitudinal axis. It is situated just dorsal to the midline. The hypothecal plates are covered with minute poroids.

Size: 27–45 μm long. Around British Isles usually at lower end of size range.

Distribution: Rare off NW Britain in Gulf Stream waters. Also recorded off the west coast of Norway, North Atlantic, eastern tropical Pacific Ocean, Straits of Florida and Gulf of Mexico.

Dinophysis nasutum (Stein) Parke & Dixon

Fig. 4G, Pl. IIc.

Parke & Dixon, 1968, p. 797.

Syn: *Phalacroma nasutum* Stein, 1883, pl. 18, figs. 1–6.
                         Entz, 1905 (1902).
Pseudophalacroma nasutum Jorgensen, 1923, p. 4, fig. 1.
                         Lebour, 1925, p. 75, figs. 20a, b.
                         Schiller, 1933, p. 55, fig. 54.
                         Tai & Skogsberg, 1934, p. 469, fig. 13.
Prodinophysis nasutum Loeblich, 1965, p. 16.

Oval or roundish in lateral view with slight constriction at girdle. Epitheca much smaller than hypotheca, domed and not obscured by the cingular lists. Characteristically the sulcus extends onto the epitheca two-thirds of the distance from the girdle to the apex, and has a rounded end with slightly raised sides. In lateral view this gives the appearance of a notch in the epithecal dome. The anterior cingular list together with the posterior list may be supported by very short ribs, the former list being horizontal or inclined anteriorly as much as 45°. The posterior cingular list resembles the anterior one but is narrower. The hypotheca is large and bears no antapical spines or protuberances. The sulcal lists are narrow and inconspicuous with no obvious ribs on the left sulcal list. Jorgensen (1923) regarded the absence of ribs as characteristic for his genus *Pseudophalacroma* but Tai & Skogsberg (1934) mention the presence of three small ribs. The sulcus extends halfway down the hypotheca. The theca is covered with coarse poroids which may contain a pore. The girdle plates have three rows of poroids arranged horizontally with the middle row each containing a pore. Megacytic stages do occur.

Size: 43–49 μm long; depth of body 38–43 μm.

Distribution: Around the British Isles. Morecambe Bay, Ireland. Elsewhere not recorded frequently but reported from the Mediterranean, west coast of Norway, between Ireland and Nova Scotia, Black Sea, Florida current.

Dinophysis norvegica Claparede & Lachmann

Fig. 4A, Pl. IIa.

Claparede & Lachmann, 1859, p. 407, pl. 20, fig. 19.
Jorgensen, 1899, p. 29, pl. 1, figs. 3–6.
Paulsen, 1907, p. 5, fig. 1; 1908, p. 14, figs. 11, 12.
Lebour, 1925, p. 79, fig. 21a.
Schiller, 1933, p. 128, figs. 122a–n.
Paulsen, 1949, p. 37, figs. 6A–F.
Solum, 1962, p. 24, fig. 14, 1–12, fig. 16, 1–12.
Balech, 1976, p. 82, figs. 2E–U.

Syn: *D. debilior* Paulsen, 1949 = *D. norvegica* var. *debilior* Pauls. 1907.

Variable in appearance but generally with low concealed epitheca and hypotheca with straight to concave sides at posterior end forming a central point when seen in lateral view. The latter feature is in contrast to the more ventral point in *D. acuta*. *D. norvegica* also differs from *D. acuta* by being generally shorter, broadest above or at the middle rather than below the middle as in *D. acuta*. The anterior cingular list is wider than the posterior list, both being projected anteriorly. They may be sculptured with fine or coarse irregular sinuous lines or reticulations. The hypotheca may be smooth in outline or coarse irregular bumps at the posterior end may be present and less frequently all round the hypothecal margin. The dorsal margin of the hypotheca is convex becoming concave to straight at the antapical end. The ventral margin is straight or convex becoming concave at the antapex. The left sulcal list is generally narrower than in *D. acuta* and curved to the right between R_2 and R_3. R_1 and R_2 are closer together than R_2 and R_3 and projected towards the anterior; R_3 is curved or straight and projects towards the posterior. The right sulcal list is short, ending between R_2 and R_3. The sulcal lists may be smooth but usually are covered with conspicuous areoli each containing a pore. The main hypothecal plates are covered with conspicuous areoles each containing a pore. The epithecal plates bear a sinuous sculpture.

Size: 48–67 μm long.

Distribution: Occurs all round the North of Britain, most abundantly in the North Sea. Its range extends south to the Suffolk coast on the east and to Anglesey on the west. Elsewhere reported from the Newfoundland Banks, from Arctic Waters, off Norway and in the Baltic, and off Wood's Hole.

Note: This species is very varied and both Paulsen (1949) and Solum (1962) have attempted to analyse it more carefully by making numerous measurements. A number of forms and varieties have been described. It is our view that the form of the hypotheca is not closely defined and the species description must allow for a fair amount of variability.

Dinophysis ovum Schutt

Fig. 3J

Schutt, 1895, pl. 1, fig. 6.
Paulsen, 1908, p. 17, fig. 16.
Lebour, 1925, p. 81, pl. 12, fig. 3.
Schiller, 1933, p. 116, fig. 109.
Abe, 1967, p. 50, figs. 10a—p.

Syn: *Dinophysis brevisulcus* Tai & Skogsberg, 1934 partim, p. 430, figs. 3a—k.

In lateral outline ovoid with a straight to slightly convex epitheca; rounded hypotheca with no protuberances. Epitheca obscured by relatively deep funnel formed by the anterior cingular list. This list is supported by ribs. The posterior cingular list is narrower and also inclined anteriorly. The cingulum is excavated on the dorsal side but not on the ventral side. The hypotheca is deepest, posterior to the middle and is covered with areoli containing a pore. The left sulcal list is relatively wide, supported by three evenly-spaced ribs and half the length of the hypotheca in extent. R_1 and R_2 are slightly concave anteriorly and R_3 concave posteriorly. It may be sculptured with areoles and is approximately the same width along its length. The right sulcal list is small, ending to the anterior of R_2.

Size: 44—58 μm; length/breadth ratio usually between 1.1 and 1.2.

Distribution: Recorded from all around coast of Britain but not in large numbers. Also from Atlantic, Mediterranean, Indian Ocean and Pacific Ocean.

Note: Some confusion occurs in delimiting this species from those which are broadly similar, i.e. *D. acuminata* and *D. recurva.* It has a lower length/breadth ratio than the former; is noticeably broader posteriorly than anteriorly, unlike the former which is more straight-sided. *D. recurva*, however, is less readily distinguished from *D. ovum.* The former is smaller, usually around 40 μm long, has a left sulcal list which between R_2 and R_3 is curved to the right. In other respects the two species are similar including their distribution.

Dinophysis pulchella (Lebour) Balech

Figs. 4E, F

Balech, 1967, p. 84

Syn: *Phalacroma pulchellum* Lebour, 1922, p. 77, pl. 11, fig. 2.
Prodinophysis pulchella (Lebour) Balech, 1944, p. 429.

Small species with conspicuous epitheca and narrow cingular lists. In lateral view subcircular with deeply incised broad girdle. In ventro-dorsal view body slightly compressed laterally but spherical in megacytic stage. The epitheca is high and rounded. The girdle lists are horizontal supported by ribs and approximately half the breadth of the girdle in width. The hypotheca is evenly rounded with no antapical spines, lists, or protuberances. The sulcus extends onto the epitheca. The left sulcal list reaches just over half-way down the hypotheca and is supported by ribs. The right sulcal list ends between R_2 and R_3. Both lists are relatively narrow. The theca is covered with fine poroids. Cell contents are pink to colourless.

Size: 21–34 µm long.

Distribution: Off SW and NE England. Also recorded from the Aegean Sea, Straits of Messina, SW Indian Ocean, off Brittany, Villefranche, in the Antarctic convergence and from Port Hacking (Australia).

Note: This species closely resembles *D. contracta* from which it differs in having ribs supporting the left sulcal list and having the sulcus extend onto the epitheca.

Dinophysis punctata Jorgensen

Fig. 3 I

Jorgensen, 1923, p. 23, fig. 28
Lebour, 1925, p. 81, pl. 12, fig. 5.
Schiller, 1933, p. 118, fig. 111.

Small species with epitheca masked and hypotheca rounded posteriorly with no protuberances. Epitheca gently convex, slightly higher dorsally. Anterior cingular list relatively wide forming a funnel over the epitheca and supported by ribs. Posterior cingular list approximately half as wide as the anterior one but similarly projected forwards. The cingulum is not excavated. The hypotheca has gently convex sides when seen in lateral view and is covered with areoles. The left sulcal list is relatively wide and supported by three evenly-shaped ribs. R_1 and R_2 project anteriorly and R_3 curves towards the posterior. The list extends three-quarters of the way down the hypotheca. The right sulcal list is short as in *D. ovum*.

Size: 28–33 µm.

Distribution: Recorded from Plymouth, the Indian Ocean, off South America, Red Sea and from Western France.

Note: This species closely resembles *D. recurva* from which it differs by being smaller, having a longer left sulcal list and by being less broad posteriorly.

Dinophysis recurva Kof. & Skogs.

Kofoid & Skogsberg 1928, p 228.
Schiller 1933, p 81. fig 105.
Wood 1968, p 51. fig 123.

Syn: *Dinophysis lenticula* Pavillard 1916, p 59, pl 3. fig 6.
                         Lebour 1925, p 81. pl 12. fig 4.

Non: *D. lenticula* Daday 1888

Cell ovate with epitheca small and completely hidden by girdle lists. General form similar to *D. punctata* (c.f) but cell is rather larger and hypotheca relatively more broad towards the posterior. Theca rather coarse and regularly punctate (or areolate). Yellow chloroplasts present.

Size: 40–50 μm long; 25–35 μm wide.

Distribution: English Channel and other areas around the British Isles; Adriatic, Mediterranean, Straits of Florida, off Japan etc.

Dinophysis rotundata Clap. & Lach.

Fig. 4 I, Pl. II f.

Claparede & Lachmann, 1859, p. 6, pl. 20, 16.
Abe, 1967, p. 57, figs. 15a–g.
Drebes, 1974, p. 116, fig. 94d.
Balech, 1976, p. 91, fig. 4, O–T.

Syn: *Phalacroma rotundatum* Kofoid & Michener, 1911, p. 290
                                   Jorgensen, 1923, p. 5, fig. 2.
                                   Lebour, 1925, p. 78, pl. 11, figs. 3a–c.
                                   Schiller, 1933, p. 67, fig. 60e.
                                   Tai & Skogsberg, 1934, p. 426, fig. 2.
       Prodinophysis rotundatum (Clap. & Lach.) Balech, 1944, p. 429.
       Dinophysis whittingae Balech, 1971, fig. 154–167.

In lateral view almost circular with low rounded epitheca which may be flattened at apex. In ventral view cell seen to be compressed but with convex sides. Both cingular lists are inclined anteriorly but the anterior one does not mask the epitheca entirely. The left sulcal list is of moderate width, widening posteriorly with three conspicuous ribs; R_1 and R_2 are closer together than R_2 and R_3. This list extends over half the length of the hypotheca. The right sulcal list is narrower than the left and is long, ending either at R_3 or to the posterior of R_3; it is sigmoid in outline with a concavity at the corresponding position of R_2.

The thecal surface is covered with poroids and scattered pores. Cell contents colourless or pinkish. Chromatophores absent. Oil globules frequently present. Megacytic stages frequent in occurrence.

Size: 36–56 µm long.

Distribution: Occurs all around British Isles especially in spring and summer, a widely distributed species. Recorded from Pacific, Atlantic and Indian Oceans. Off Norway, Greenland, Nova Scotia, from the Baltic, Mediterranean, Black Sea, Gulf of Siam, Barents Sea, Inland Sea of Japan and Arabian Sea.

Note: *Phalacroma rudgei* and *Dinophysis whittingae* are considered by Sournia (1973) to be synonyms, and may represent the megacytic stage of *Dinophysis rotundata.*

SINOPHYSIS Nie & Wang

Nie & Wang, 1944, p. 146.

Syn: *Phalacroma* Stein partim
Dinophysis Ehrenberg partim
Thecadinium Kofoid & Skogsberg partim

Body more or less ellipsoidal in lateral view, compressed laterally. Epicone very small and tilted to the dorsal side; girdle inclined towards dorsal; upper list wide, lower list narrow. Sulcus extending to about two-thirds length of cell with slight list on left border. Cell covered with thin plates having delicate reticulations. Cytoplasm colourless (no chloroplasts) with nucleus at the posterior end of the cell.

Type species: *Sinophysis microcephalus* Nie & Wang, 1944.

Sinophysis ebriolum (Herdman) Balech

Fig. 3 D

Balech, 1956, p. 21, figs. 9–22.

Syn: *Phalacroma ebriola* Herdman, 1924, p. 79, fig. 24.
P. ebriolum Lebour, 1925, p. 77, fig. 20C.
Dinophysis ebriola Herdman, 1925, p. 82.
Thecadinium ebriolum Kofoid & Skogsberg, 1928, p. 32.
T. ebriolum Schiller, 1933, p. 51, fig. 50.

Cell flattened laterally, roughly ellipsoidal. Epicone very small button-like, surrounded by a wide collar formed by the anterior girdle list; hypocone sack-shaped with a flattened top. Girdle deep, tilted to be lowest on dorsal side, bounded by lists. Sulcus about half the length of the cell, bordered by

2 plates, girdle 4 plates, hypotheca of 2 main convex plates which cover the sides of the cell, and a narrow plate is found in the ventral area. The cell is colourless but contains coloured food bodies; two large vacuoles in the anterior part; ellipsoidal nucleus in the lower part of the hypocone.

Size: 32–43 μm long; 25–31 μm wide.

Distribution: Sand at Port Erin, Isle of Man; sand in Brittany; sand at Boulogne, English Channel; Wood's Hole, U.S.A.

THECADINIUM Kof. & Skogs

Kofoid & Skogsberg, 1928, p. 32.

Syn: *Amphidinium* Claparede & Lachman partim
       Phalacroma Stein partim

Cell laterally flattened; hypocone reduced to about one quarter of the body length. Girdle anterior, sulcus on the narrow ventral side and may extend beyond antapex. Hypotheca mainly covered by two large lateral plates. Cell with or without chloroplasts. Generally found in wet sand.

Type Species: *Thecadinium petasatum* (Herdman) Kofoid & Skogsberg.

Note: The type species was originally described as one of the group of laterally flattened *Amphidinium* species. Lebour (1925) transferred it to *Phalacroma* because of the cell proportions and large hypothecal plates. Kofoid & Skogsberg (1928) created the new genus *Thecadinium* and clearly regarded it as a member of the order Dinophysiales. Later Balech (1956) who described a number of new species (some of which are now transferred to *Amphidiniopsis*) thought that the genus belonged in the order *Peridiniales*. However, it is clear that the reduced genus, as here presented, has the dinophysoid type of organisation and it is now returned to the Dinophysiales.

Thecadinium petasatum (Herdman) Kof. & Skogs.

Figs. 5A–C

Kofoid & Skogsberg, 1928, p. 32–33.

Syn: *Amphidinium kofoidi* var. *petasatum* Herdman, 1922, p. 26, fig. 3
       Phalacroma kofoidi Lebour, 1925, p. 77, pl. 11, fig. 1.
       Thecadinium kofoidi Schiller, 1933, p. 51, fig. 51.

Cell compressed laterally, from side view ± ellipsoidal with a rather flattened anterior end. Epicone less than one quarter of the cell length, triangular ventrally, rounded dorsally. Hypocone rounded at the antapex in lateral view, bluntly pointed when seen (rarely) in front view. Girdle angled on the ventral side to meet the sulcus half way down the cell. Sulcus short, on hypocone only. Flagella arising from junction of girdle and sulcus, longitudinal flagellum 1–1½ times cell length. Ovoid nucleus situated in lower part of cell; numerous radiating chloroplasts arising from the central pyrenoid. Cell covered with a firm theca. The hypocone is covered by two large lateral plates with scattered pores plus a sulcal plate structure. The epicone has approximately 4 plates.

Size: 25–33 μm long; lateral breadth 20–26 μm.

Distribution: Found at Ramsey and its original site at Port Erin, Isle of Man; Galloway, Scotland; and also reported from Brittany.

Thecadinium semilunatum (Herdman) Comb. nov.

Fig. 5 D

Syn: *Amphidinium semilunatun* Herdman, 1923, p. 59, fig. 7.
         A. semilunatum Lebour, 1925, p. 33, fig. 9.
         A. semilunatum Schiller, 1933, p. 315, fig. 317.
         Thecadinium inclinatum Balech, 1956, p. 40, figs. 34–37.

Body flattened laterally and shaped as an irregular ellipsoid from side view. Epicone about one quarter the length of the cell, rounded, but sometimes flattened, extended onto the ventral side of the cell. Hypocone sac-shaped, rather straight on the dorsal side, rounded on the ventral, and squarish to rounded antapically. The girdle is high on the dorsal side and lowest ventrally, the sulcus is deep and extends from the girdle past the antapex to the lower dorsal corner of the cell, thus causing a deep furrow in the cell if viewed from the ventral side. Flagella arise from junction of sulcus and girdle; longitudinal flagellum about 1½ times body length. The nucleus is situated in the lower ventral region and the cytoplasm is colourless to yellowish with a glassy appearance, it also contains refractile granules which may be food vacuoles. There are no chloroplasts. The cell is covered by a delicate theca the plates of which have not been analysed but in *T. inclinatum* Balech (1956) reports three triangular plates in the sulcal region.

Size: 50–64 μm long; 30–48 μm in lateral breadth.

Distribution: Originally described from the sandy beach at Port Erin, recently found on a beach at Folkestone (Kent).

Note: Although it has not been possible to analyse the thecal plates of *T. semilunatum* it would seem, from shape and general characteristics, that Balech's *T. inclinatum* is the same species and it is here brought into synonymy. *Amphidinium sulcatum* Kofoid (1907a) also has the general characteristics of this species although it possesses yellow chloroplasts. Further studies are needed to see if this should be transferred to *Thecadinium*.

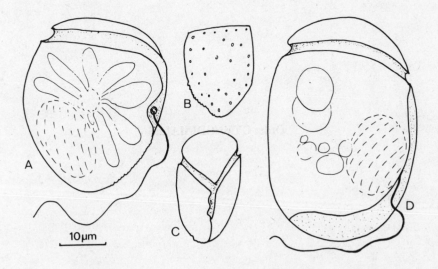

Fig. 5   THECADINIUM
A. *T. petasatum* right side B. *T. petasatum* main hypothecal plate from the left side (not to scale) C. *T. petasatum* ventral view (not to scale)
D. *T. semilunatum*

Order GYMNODINIALES

Family Gymnodiniaceae

AMPHIDINIUM Claparede & Lachman

Claparede & Lachman, 1859, p. 410, pl. 20, figs. 9–12.

A genus of naked dinoflagellates distinguished by having a much reduced epicone and the girdle situated near the anterior end of the cell. The sulcus is usually only found on the hypocone but may run from apex to antapex. The cell is typically dorso-ventrally flattened but a few species have rounded cells. (The laterally flattened species in Lebour (1925) and other early works are now placed in other genera). The flagella are generally inserted a short distance apart, often there is a 'bridge' between the two pores. Chloroplasts may be present and then there is usually (but not in *A. britannicum*) a single large pyrenoid. Species without chloroplasts may contain coloured granules. Nutrition is autotrophic or heterotrophic and at least one species is phagotrophic. The nucleus is usually situated in the posterior end of the cell and varies from sphaerical to U-shaped. One or more pusules may be present, associated with the flagellar canals. The cell is covered by a delicate membranous theca (see Dodge & Crawford, 1968; 1970, for details) and consequently the cells usually distort or change shape on fixation.

Type species: *A. operculatum* Claparede & Lachman, 1859.

Ecology: The majority of species found around the British Isles are littoral organisms living in rock pools, salt marshes, and among the sand grains on beaches. In the latter case the *Amphidinium* may be present in sufficient numbers to colour the beach green or brown at low tide and the organisms migrate from the surface down into the sand when the tide comes in. The main species are probably distributed all around the British Isles. Some species are very robust and will grow readily in rough cultures made from shore collections. Many are extremely tolerant of salinity variations and some supposedly freshwater species can be found on the shore.

The present account deals with 17 species which have all been observed in recent years. The latest check-list (Parke & Dodge in Parke & Dixon, 1976) does, however, include 55 *Amphidinium* species. Of these 26 have only been reported from estuarine sites on the coasts of Belgium (Conrad, 1926, 1939; Conrad & Kufferath, 1954) and 10 species are of doubtful provenance.

There is considerable taxonomic confusion in this genus which mainly results from the difficulty of making accurate descriptions of cells which are one minute fast-moving and the next are changing shape as a result of death or fixation. Taylor (1971a) attempted to clear up some of the confusion surrounding the species *A. klebsii* and *A. carterae* but it would appear that there is much scope for bringing into synonymy many of the described taxa. Species such as *A. britannicum* appear to show an intermediate form with the genus *Gyrodinium* and others may be near to *Gymnodinium*.

KEY TO THE SPECIES OF *AMPHIDINIUM*

1a	Cell containing chloroplasts	2
b	Cell not containing chloroplasts although yellowish or bluish pigmented granules may be present	9
2a	Epicone reduced to a small straight or bent peg or triangular in outline	3
b	Epicone substantial, rounded, whole width of cell	6
3a	Epicone straight, a rounded peg	*A. lacustre*
b	Epicone bent and/or triangular	4
4a	Epicone triangular with slight bend	*A. herdmanae*
b	Epicone peg or beak-like	5
5a	Chloroplasts radially arranged from central pyrenoid	*A. operculatum*
b	Chloroplasts forming a peripheral network	*A. carterae*
c	Chloroplasts numerous, discoid	*A. ovoideum*
6a	Anterior of cell notched	*A. corpulentum*
b	Anterior of cell entire	7
7a	Epicone asymmetrical, cell large	*A. britannicum*
b	Epicone symmetrical	8
8a	Girdle more or less horizontal, cell elliptical	*A. ovum*
b	Girdle V-shaped on ventral side of cell, rounded	*A. globosum*
9a	Cell rounded	10
b	Cell dorso-ventrally flattened	11
10a	Cell spindle-shaped, pointed at both ends	*A. sphenoides*
b	Cell ±symmetrical, rounded at anterior, pointed at posterior end	*A. mannanini*

11a	Antapex deeply notched	*A. bipes*
b	Antapex slightly notched or entire	12
12a	Epicone symmetrical, entire	13
b	Epicone not symmetrical, may be notched	14
13a	Epicone triangular, antapex sharply rounded	*A. crassum*
b	Epicone rounded and flattened, antapex rounded	*A. latum*
c	Epicone rounded, antapex squarish to broadly rounded	*A. flexum*
14a	Epicone small, triangular, cell slender	*A. scissum*
b	Epicone large, rounded, cell broad	*A. pellucidum*

Amphidinium bipes Herdman

Fig. 7 C

Herdman, 1924, p. 78, fig. 19.
Lebour, 1925, p. 29, fig. 8f.
Schiller, 1933, p. 280, fig. 267.

Cell with rather squarish anterior end and distinctly bifurcated posterior end. Somewhat flattened dorso-ventrally. Epicone triangular, not as wide as cell and slightly grooved by sulcus. Girdle more or less symmetrical and V-shaped on ventral surface. Hypocone of two rounded lobes with broad and shallow sulcus. Longitudinal flagellum free from its point of origin and nearly 2 x cell length (often it is turned forwards). Nucleus situated near centre of the cell, on the right side. Refractile granules may be present, particularly at posterior end.

Size: about 30 μm long; 22 μm wide.

Distribution: Sand at Port Erin; Isle of wight, Sussex, Kent, York-shire and Galloway.

Amphidinium crassum Lohmann

Fig. 6 A

Lohmann, 1908, p. 252, pl. 17, fig. 16.
Lebour, 1925, p. 31, pl. 3, fig. 2.
Schiller, 1933, p. 283, fig. 272.

Syn: *A. phaeocysticola* Lebour, 1925, p. 31, pl. 3, fig. 3.

Body elongate elliptical in ventral view, nearly circular in cross-section. Epicone small, a symmetrical cone with the sulcus reaching to the apex. Antapex rounded or pointed, possibly showing longitudinal striations. Girdle wide, not displaced, sulcus narrow and straight. Transverse flagellum encircling the cell, posterior flagellum approximately 2 x length of cell. Nucleus large, spherical, posterior in position. No chloroplasts but brown bodies (food vacuoles?) may be present.

Size: 23–42 μm long; 11–24 μm wide.

Distribution: Found in Plymouth Sound and on most sandy beaches which have been sampled. Also reported from Belgium, Baltic Sea, eastern U.S.A., Adriatic Sea.

Note: *A. phaeocysticola* found by Lebour amongst blooms of *Phaeocystis* (which it was apparently ingesting) is almost certainly a large form of *A. crassum*.

Amphidinium flexum Herdman

Fig. 6 B

Herdman, 1923, p. 59, fig. 3.
Lebour, 1925, p. 27, fig. 9a.
Schiller, 1933, p. 292, fig. 284.

Body rounded-rectangular in ventral view, dorso-ventrally flattened. Epicone broad and low with a rounded apex, slightly grooved by the sulcus. Hypocone twisted with the left side displaced backwards and the sulcus running to the antapex which is distinctly bilobed. Girdle symmetrical, not very broad. Transverse flagellum encircling cell, longitudinal flagellum about 2 x body length. Nucleus posterior in position; cytoplasm uncoloured but usually contains numerous small yellow bodies/granules.

Size: 42–60 μm long; 28–40 μm wide.

Distribution: Sand at Port Erin; Galloway; Pembrokeshire; English Channel; Orkneys; Belgian coast.

Note: It would seem possible that this and *A. pellucidum* are forms of the same species but this requires investigation.

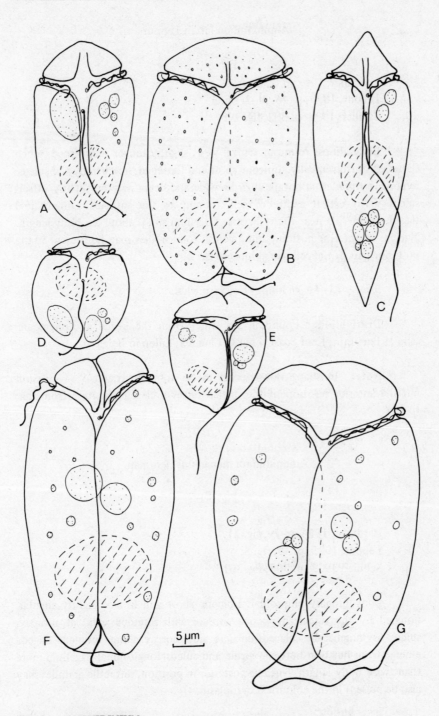

Fig. 6 AMPHIDINIUM
A. *A. crassum* B. *A. flexum* (cf. Herdman) C. *A. sphenoides* (after Hulburt 1957) D. *A. latum* E. *A. manannini* F. *A. scissum* G. *A. pellucidum*

Amphidinium latum Lebour

Fig. 6 D

Lebour, 1925, p. 26, pl. 2, fig. 3.
Schiller, 1933, p. 302, fig. 298.

Body almost round in ventral view, slightly longer than broad, much flattened dorso-ventrally. Epicone short and flattened, notched by the sulcus; hypocone rounded at antapex; girdle wide and deep, may be slightly offset; sulcus narrowest at centre of cell, running to the antapex. Longitudinal flagellum emerges near the transverse flagellum and is about 2 x body length. Nucleus in anterior position and greenish food bodies may be present in the posterior part of the cell. No chloroplasts.

Size: 13—19 µm long; 11—17 µm wide.

Distribution: Found in sea water from the beach at Cullercoats; sand at Port Erin; Kent coast, English Channel; Lilloo in Belgium.

Note: In shape this appears to be a small version of *A. crassum* but the anterior position of the nucleus gives a clear distinction from that species.

Amphidinium manannini Herdman

Fig. 6 E

Herdman, 1924, p. 79, fig. 21.
Lebour, 1925, p. 34, fig. 8, o.
Schiller, 1933, p. 303, fig. 301.

Body broadly ellipsoidal. Epicone short and broad, slightly angular, notched by the sulcus; hypocone cordate with rounded point at antapex; sulcus asymmetrically placed on left side not reaching antapex. flagella emerge from junction between girdle and sulcus, longitudinal flagellum more than twice body length. Nucleus posterior in position, refractile granules may also be present in the colourless cytoplasm.

Size: about 20 µm.

Distribution: Sand at Port Erin, Isle of Man, and Galloway (Scotland); brackish water in Belgium.

Amphidinium pellucidum Herdman

Fig. 6 G

Herdman, 1922, p. 27, fig. 7.
Lebour, 1925, p. 28, fig. 8d.
Schiller, 1933, p. 309, fig. 309.

Body broadly elliptical, flattened dorso-ventrally. Epicone broad
with rounded apex, sulcus narrow almost extending to apex. Hypocone
broadly sac-shaped with rounded and notched posterior, sulcus extending
to antapex, but obscured by projections from either side. Girdle slightly
asymmetric, V-shaped on ventral side of cell. Transverse flagellum encircling
cell and longitudinal flagellum lying in sulcus (N.B. perhaps it emerges two-
thirds of way down cell) and extending almost one cell length. Nucleus
either near centre of cell (Herdman) or in posterior end (Conrad & Kufferath).
Cytoplasm colourless and 'glassy' but may contain red coloured bodies
(? food vacuoles), particularly behind the sulcus.

Size: 50–60 µm long; ca. 35 µm wide.

Distribution;   Sand at Port Erin; Hebrides; brackish water in Belgium;
sand in Brittany.

Amphidinium scissum Kofoid & Swezy

Fig. 6 F

Kofoid & Swezy, 1921, p. 150, pl. 2, fig. 22, fig. U1
Lebour, 1925, p. 26, pl. 2, fig. 4.
Schiller, 1933, p. 319, fig. 316.

Syn: *A. scissoides* Lebour, 1925, p. 30, pl. 3, fig. 1.

Body elongate oval, flattened dorso-ventrally. Epicone flattened
and asymmetric, extended to the left, cut into two portions by the sulcus.
Girdle slightly displaced and forming a V on ventral side. Hypocone with
rounded antapex, sulcus broadening before reaching this. Transverse
flagellum encircling cell from origin at junction of girdle and sulcus. Longitudinal flagellum emerging three quarters of the way down the cell but only
extending about half cell length beyond it. Nucleus very large, occupying
posterior end of cell. Cytoplasm colourless or greenish with round refractive
bodies. Cell surface may be striated.

Size: 50–60 µm long; ca. 30 µm wide.

Distribution: Sandy beaches at Port Erin; the Wash, Kent coast,
Galloway, California.

Note: *A. scissoides* Lebour (1925) of which only one cell was seen by Lebour was probably a swollen cell which had been washed out to sea.

Amphidinium sphenoides Wulff

Fig. 6 C

Wulff, 1916, p. 105, pl. 1, fig. 9.
Lebour, 1925, p. 30, fig. 81.
Schiller, 1933, p. 315, fig. 318.
Hulburt, 1957, p. 203, figs. 4, 9, 13.

Body spindle-shaped, tapered posteriorly to a sharp point and anteriorly to a shorter point. Epicone small and triangular, tilted forwards in side view, slightly grooved by sulcus. Girdle more or less horizontal, wide dorsally and narrow ventrally. Hypocone very elongate, more or less symmetrical in all views, sulcus extending just over half-way to the antapex. Flagella inserted near junction of sulcus and girdle; transverse flagellum encircling cell, posterior flagellum short, scarcely extending beyond antapex. Nucleus situated just below mid-point of cell; no chloroplasts but numerous granules present.

Size: 35–50 (65) μm long; 12–15+ μm wide.

Distribution: Sandy beach Port Erin; Plymouth Sound; Helgoland; Barents Sea; Woods Hole, U.S.A.

Note: Lebour (1925) pointed out the similarity between *A. sphenoides* and the organism which she had described as *Gymnodinium filum.* Drebes (1974) described a dinoflagellate which parasitises the diatom *Chaetoceros.* The motile stages of this parasite appear to be identical with *A. sphenoides* and this would therefore appear to be the only known parasitic member of the genus.

Amphidinium britannicum (Herdman) Lebour

Fig. 7 E

Lebour, 1925, p. 27, pl. 2, figs. 5–6.
Schiller, 1933, p. 280, fig. 268.

Syn: *A. asymmetricum* var. *britannicum* Herdman, 1922, p. 21, fig. 5.

Body asymmetrically oval, longest on the left; dorso-ventrally flattened. Apex and antapex rounded. Epicone large and distinctly asymmetrical; girdle in a steep spiral. Hypocone long, sac-shaped but may be notched by sulcus; sulcus only on hypocone and concealed by a flap. Transverse flagellum encircles cell; longitudinal flagellum arises near posterior end and is just over cell length. Nucleus spherical and central in position.

Chloroplasts numerous, radiating from near the centre of the cell. No p
Cell surface smooth.

Size: 37–60 μm long; c. 30 μm wide.

Distribution: abundant on the sandy beach at Port Erin in the spring and summer; also found on sandy beaches along south coast of England, Yorkshire and SW and E Scotland.

Note: In its overall shape and the position of the girdle this species is clearly intermediate between *Amphidinium* and *Gyrodinium*. In its dorso-ventral flattening, ecological niche and chloroplast type it appears to be closest to *Amphidinium*.

Amphininium carterae Hulburt

Fig. 7 J

Hulburt, 1957, p. 199, pl. 1, fig. 1.
Taylor, 1971, p. 131, fig. 2.

Syn: *A. klebsii* Carter, 1937, p. 58, pl. 8, figs. 12–15.

Body broadly oval in ventral view; dorso-ventrally flattened. Epicone small, asymmetric, directed to left; cresent shaped in ventral view, button-shaped in lateral view. Girdle encircles epicone except for small area on left side, being low and V-shaped ventrally and high dorsally. Hypocone rounded at apex with sulcus running from just anterior to termination of girdle, along right side of hypocone. Flagellar pores separated by bridge of girdle, longitudinal flagellum 1.5 x cell length. Nucleus ovoid, in posterior end of cell. Only one chloroplast, peripherally arranged and subtending a large anterior pyrenoid.

Size: 11–24 μm long; 6–17 μm wide (Hulburt)

Distribution: An extremely common species of rock pools, wet sand areas and salt marshes. It is a very hardy organism and will readily grow in the laboratory from enrichment cultures. Present all round the British Isles and reported from many parts of the world (see Taylor, 1971).

Amphidinium corpulentum Kofoid & Swezy

Fig. 7A

Kofoid & Swezy, 1921, p. 134, pl. 1 fig. 11, fig. U6, 13
Schiller, 1933, p. 282, fig. 271.

Body sac shaped, dorso-ventrally flattened. Epicone broad and about one-fifth of the total cell length, rounded with distinct apical notch caused by sulcal groove reaching apex. Antapex flattened or broadly rounded and may be slightly notched by sulcus. Girdle V-shaped, lowest on ventral side, lower part of sulcus overhung by flaps. Both flagella arise near the junction of girdle and sulcus and the longitudinal flagellum protrudes about ¾ the length of the body. Nucleus ovoid, in posterior end of cell; cell packed with numerous discoid chloroplasts and a red spot present in the mid hypocone.

Size: 35 μm long, 18 μm wide (in California 46–54 long; 30–34 μm wide).

Distribution: Sandy beach on Romney Marsh, Kent; Beach sand at La Jolla, California.

Amphidinium globosum Schroder

Fig. 7 F

Schroder, 1911, p. 651, fig. 16.
Schiller, 1933, p. 294. fig. 287.

Cell almost globular with apex and antapex rounded. Epicone symmetrical; girdle forming a shallow V on the ventral side and straight on dorsal side. Hypocone widest at the middle of the cell, only slightly grooved by the sulcus. Several chloroplasts present.

Size: 8–16 μm long; 7–14 μm wide.

Distribution: Pool on a sandy beach at Littlehampton, Sussex; previously only known from the Adriatic Sea.

Additional species: These have been reported from around the British Isles but have not been found during the present survey:

A. *pelagicum* Lebour, 1925, p. 32, pl. 3, fig. 4.
         Schiller, 1933, p. 308, fig. 308.
A. *vitreum* Herdman, 1924, p. 79, fig. 22.
         Lebour, 1925, p. 34, fig. 8m.
         Schiller, 1933, p. 322, fig. 327.

Amphidinium herdmanae Kofoid & Swezy

Fig. 7 H

Kofoid & Swezy, 1921, p. 143, fig. U2.
Lebour, 1925, p. 23, pl. 1, fig. 2.
Schiller, 1933, p. 294, fig. 288.

Syn: A. *operculatum* Herdman, 1911, p. 554.

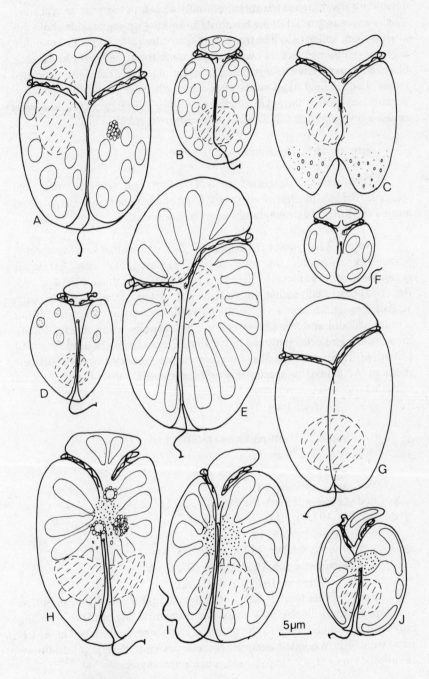

Fig. 7   AMPHIDINIUM
A. *A. corpulentum* B. *A. ovoideum* C. *A. bipes* D. *A. lacustre*
E. *A. britanicum* F. *A. globosum* G. *A. ovum* H. *A. herdamanae*
I. *A. operculatum* J. *A. carterae*

Cell ellipsoidal with rounded apex and antapex. Much flattened dorso-ventrally. Epicone triangular, centrally placed, not as wide as hypocone, usually extending to left. Girdle starting and ending on ventral side about one quarter along length of cell but much higher on dorsal side. Hypocone rounded, may be slightly notched by sulcus which runs from the girdle to antapex. Flagellar pores situated near junction of girdle and sulcus, each with associated pusule. Longitudinal flagellum 1—1½ x cell length. Nucleus posterior, crescent-shaped. Numerous chloroplasts, radiating from a central pyrenoid. Red spots occasionally present in cell together with numerous starch grains.

Size: 25—50 μm long; 17—25 μm wide.

Distribution: Abundant in sand at Port Erin where it forms conspicuous stained patches; Cullercoats (Northumberland) and many places around the British Isles. A typical species of damp sand.

Note: Like most dinoflagellates this is a rather variable species, particularly as regards the form of the epicone, which may be almost symmetrical or considerably deflected to the left. Herdman (1924, p. 76, figs. 2—5) described another *Amphidinium, A. testudo,* from the sandy beach at Port Erin which appears very similar to *A. herdmanae* except that it has a much reduced and withdrawn epicone and appears to be surrounded by some form of pellicle. Herdman noted that it was variable both in form and colour. Although as yet we have no information on encystment or resistant stages of Amphidinium species it would seem possible that *A. testudo* is an encysting form of *A. herdmanae.*

Amphidinium lacustre Stein

Fig. 7 D

Stein, 1883, p. 15, pl. 17, figs. 21—30.
Schiller, 1933, p. 300, fig. 294.

Syn: *A. schroderi,* Schiller, 1928, p. 134, fig. 10.
       A. lacustriforme Schiller, 1928, p. 132, fig. 6.

Body ovoidal to chordate, not flattened. Epicone tiny, a small button raised on a wide girdle. Girdle V-shaped on the ventral side of the cell, leading into the sulcus which runs to the antapex. Hypocone heart-shaped in ventral view, with slightly pointed antapex. Nucleus posterior. Several golden-brown chloroplasts present, a large pyrenoid, and numerous granules.

Size: 15—23 μm long; 10—18 μm wide.

Distribution: Prawle Point, Devon; Coast of Belgium; Adriatic sea; many other brackish and fresh-water sites. This species is said to be essentially one of fresh water but it obviously can grow in brackish situations.

Amphidinium operculatum Claparede & Lachman

Fig. 7 I

Claparede & Lachmann, 1858–9, p. 410, pl. 20, figs. 9–10.
Lebour, 1925, p. 22, fig. 8a,
Schiller, 1933, p. 288, fig. 304.

Syn: *A. steinii* Lemmermann, 1910, p. 580, 616, figs. 1–7.
A. klebsii Kofoid & Swezy, 1921, p. 144, fig. U14.
A. massartii Biecheler, 1952 p. 24, figs. 4, 5.
A. rhynchocephalum Anissimowa, 1926, p. 192, figs. 1–7.
A. wislouchi Hulburt, 1957, p. 199, pl. 1, fig. 2.
A. hofleri Schiller & Diskus, 1955, p. 1, figs. 1–2.

Body broadly ellipsoidal in ventral view, dorso-ventrally flattened. Epicone small and slightly bent to the left, only one third the width of the cell. Girdle high on dorsal side of cell, V-shaped on ventral side. Hypocone broadly rounded with the sulcus running its whole length. The two flagella arise near each other adjacent to the junction of girdle and sulcus. Nucleus ovoid, posterior in position. Chloroplasts numerous, radially arranged from a central pyrenoid. Orange globules may be present. Surface of cell may be faintly striate.

Size: 15–50 μm long; 10–30 μm wide.

Distribution: Wet beaches around most of the British Isles; also known from the Norwegian coast; Naples, south of France; New England; Baltic Sea; Belgium, etc.

Note: This species has been the cause of considerable taxonomic confusion. Some workers have described as new species very minor variants of the rather imprecisely figured type species while others such as Lebour (1925), Skuja (1939) and Norris (1967) have been unhappy about the validity of many of the newer species. Here, several of the *operculatum*-like organisms are recombined with the type species. In terms of shape and size there is still a problem of distinguishing this from *A. carterae* for the only difference is in the arrangement of the chloroplast as pointed out by Taylor (1971a). It is notable that *A. operculatum* was frequently reported in the literature until *A. klebsii* was described in a much more precise manner. Since that time, *klebsii* has been reported many times and *operculatum* but rarely.

Amphidinium ovoideum Lemmermann

Fig. 7 B

Lemmermann, 1900, p. 155.

Lebour, 1925, p. 23, fig. 8c.
Schiller, 1933, p. 306, fig. 305.

Syn: *Prorocentrum ovoideum* Lemmermann, 1896, p. 147, figs. 1–3.

Body broadly oval, dorso-ventrally flattened. Epicone small, tongue-shaped, deflected to the left. Hypocone broadly rounded with sulcus running only two thirds of its length. Girdle narrow surrounding base of epicone. Nucleus in lower half of hypocone, cell filled with numerous green or brownish chloroplasts, refractile granules also present.

Size: 17–23 μm long; 14–19 μm wide.

Distribution: Salt-marsh on Solway Firth; brackish water in the Baltic Sea; Belgian coast.

Amphidinium ovum Herdman

Fig. 7 G

Herdman, 1924, p. 78, fig. 25.
Lebour, 1925, p. 29, fig. 8g.
Schiller, 1933, p. 307, fig. 306.

Body ellipsoidal, flattened dorso-ventrally. Epicone broad, slightly asymmetrical, rounded at apex. Girdle slightly offset, being lower on the right side. Hypocone slightly notched by the sulcus which runs from the girdle to the antapex (overlapped by left margin) and is said (Herdman) to continue as a narrow furrow up the dorsal side of the cell. Longitudinal flagellum twice as long as cell. Nucleus ovoid, posterior; chloroplasts numerous, radiating from the centre.

Size: 24–36 μm long; 16–26 μm wide.

Distribution: Sand at Port Erin; beaches in Pembrokeshire and S.W. Scotland (Galloway); Kent.

Note: In many respects this species appears closely related to *A. britannicum* and it could be regarded as intermediate between that species and *A. herdmanae*.

COCHLODINIUM Schutt

Schutt, 1896, p.5, figs. 1,7.

Syn: *Gymnodinium* Stein partim

Naked dinoflagellates in which the body is twisted at least 1.5 turns. The girdle forms a descending left-handed spiral of 1.5 turns or more and has a considerable displacement of the two ends. The sulcus is also twisted around the cell by 0.5 turns or more. The fairly large nucleus is situated in a central or posterior position and the cell may be colourless or contain pigmentation. A few species contain chloroplasts. Pusules are found associated with the flagellar pores. Frequently found encysted in a transparent (? mucilaginous) ovoid structure. Some species tend towards a colonial habit.

Type species: *C. strangulatum* Schutt, 1896.

Note: This genus appears to be in urgent need of a complete revision, particularly with a view to discovering how fixed is the degree of torsion of the cell, but this is hindered by the scarcity of specimens.

Cochlodinium achromaticum Lebour

Fig. 8

Lebour, 1925, p.63, pl. 9, fig. 6.
Schiller, 1933, p. 511, fig. 540.

Cell a rounded ovoid, with rounded apex and flattened to depressed antapex. Girdle deeply incised in a descending left spiral of two turns. The sulcus commences in a loop near the apex, makes one turn between the start and finish of the girdle, and finishes in an antapical notch. Cytoplasm clear; nucleus posterior; food vacuoles and refractive bodies present in cytoplasm.

Size: length 43 μm; breadth c 30 μm

Distribution: Only reported from outside Plymouth Sound.

Cochlodinium brandtii Wulff

Fig. 8H

Wulff, 1916, p. 108, fig. 17a, b.
Lebour, 1925,p.65, pl. 9, fig. 8.
Schiller, 1933, p. 514, fig. 554.

Syn: *C. augustum* Kofoid & Swezy, 1921, p. 354, fig. HH 15.

Cell spindle-shaped with rounded ends. Girdle deeply incised making about 4 turns with a total displacement of over 0.75 of cell length. Sulcus only just entering the epicone but makes over 3 turns between girdle ends.

Size: 50–110 μm long.

Distribution: English Channel off Plymouth.

Cochlodinium helicoides Lebour

Fig. 8G

Lebour, 1925, p.62, pl.9, fig. 2.
Schiller, 1933, p. 528, fig. 558.

Syn: *C. helix* Schutt, 1895, pl. 22, fig. 77(5).
  C. helix Kofoid & Swezy, 1921, p. 370, fig. HH 8.

Body asymmetrically sub-oval, left posterior slightly extended. Girdle making 1.5 turns, deeply incised; sulcus making about 0.6 turns, making slight notch at antapex. Large nucleus in a central position; chromatophores present and cell a yellow colour. A thin hyaline cyst may surround the cell.

Size: 29–54 μm long; 24–30 μm wide.

Distribution: Found off Plymouth and in the Irish Sea; also reported from Brittany, the Mediterranean, Pacific Ocean.

Note: This species was created by Lebour for the organism which had been called *C. helix* by Kofoid & Swezy and which figured in one only of Schutt's figures of 1895.

Cochlodinium helix (Pouchet) Lemmermann

Fig. 8E

Lemmermann, 1899, p. 360.
Lebour, 1925, p.62, pl. 9, fig. 3.
Schiller, 1933, p. 529, fig. 559.

Syn: *Gymnodinium helix* Pouchet, 1887, p. 94–96, fig. 1.
  G. helix Schutt, 1895, pl. 22, fig. 77 (1–4).

Non: *Cochlodinium helix* Kofoid & Swezy, 1921, p. 370.

Cell ± oval in outline, distinctly coiled due to the deep incision of girdle and sulcus. Epicone rounded or conical; girdle making two complete

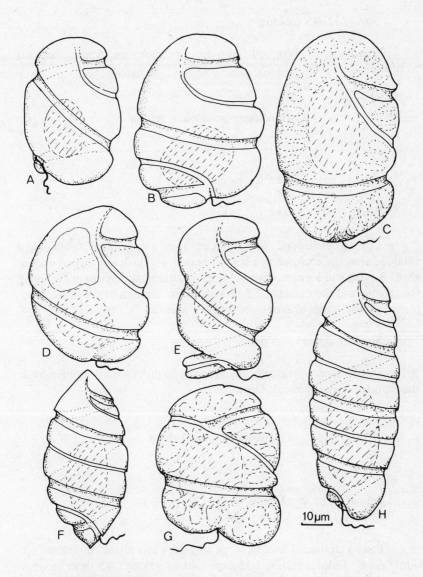

Fig. 8 COCHLODINIUM
A. *C. pupa* B. *C. achromaticum* C. *C. schuettii* D. *C. vinctum* E. *C. helix*
F. *C. pulchellum* G. *C. helicoides* H. *C. brandtii*

turns; hypocone asymmetric with a posterior process on the left side onto which the girdle runs. Sulcus just invading the epicone, then making one complete turn. Flagella short. Nucleus central; cytoplasm yellow (? chloroplasts). Sometimes enclosed in a transparent cyst.

Size: 32–45 μm long

Distribution: Plymouth; also reported from the Atlantic and off E. Australia.

Cochlodinium pulchellum Lebour

Fig. 8F

Lebour, 1917, p. 197, fig. 14.
Lebour, 1925, p. 65, pl. 9, fig. 7.
Schiller, 1933, p. 534. fig. 564.

Cell spindle-shaped with pointed apex and pointed to flattened antapex; generally enclosed in a colourless cyst. Girdle deeply incised making about 3 turns and a displacement of 0.75 of total length. Sulcus just entering the epicone and then making 2.5 turns between the ends of the girdle. Cytoplasm colourless, pale blue or green; nucleus posterior.

Size: length 35–50 μm

Distribution: Originally found in the English Channel off Plymouth; also reported from the Pacific Ocean.

Cochlodinium pupa Lebour

Fig. 8A

Lebour, 1925, p. 63, pl. 9, fig. 4.
Schiller, 1933, p. 535, fig. 565.

Body a grooved ellipsoidal shape with apex and antapex both rounded. Girdle deeply incised, making 1.75 turns; sulcus making 0.75 turns between start and finish of the girdle, also running onto the epicone and notching the antapex. Cytoplasm colourless and pellucid; refractive bodies present.

Size: length 36–40 μm; breadth c 26 μm.

Distribution: Originally reported from beyond Plymouth Sound; recently found in the Irish Sea.

Cochlodinium schuetti Kofoid & Swezy

Fig. 8C

Kofoid & Swezy, 1921, p. 380, fig. HH 2.
Lebour, 1925, p. 61, pl. 9, fig. 1.
Schiller, 1933, p. 537, fig. 568.

Syn: *Gymnodinium helix* Schutt, 1895, pl. 24, fig. 77(6) pro parte.

Body more or less ellipsoidal about 1.5 times as long as broad. Epicone rounded, sometimes with a flattened apex; hypocone rounded. Girdle making 1.5 turns, deeply incised; sulcus running a short way onto the epicone, making a twist of 0.5 turns. Nucleus posterior to central in position, cytoplasm yellow containing scattered greenish droplets, probably contains chloroplasts. Pusules may be present. Cell normally enclosed in a hyaline cyst.

Size: 50–74 μm long; c 35 μm wide.

Distribution: Plymouth, Firth of Lorne (Scotland); Pacific Ocean.

Cochlodinium vinctum Kofoid & Swezy

Fig. 8D

Kofoid & Swezy, 1921, p. 384, fig. HH. 3, pl. 2, fig. 15.
Lebour, 1925, p. 63, pl. 9, fig. 5.
Schiller, 1933, p. 541, fig. 572.

A small species. Body a rounded ovoid, arched on the dorsal side. Apex rounded with small pimple, antapex rounded but slightly cleft by the sulcus. Girdle a descending spiral of 1.5–1.7 turns with the two ends displaced about 0.6 of total cell length. Sulcus starting as an apical loop, rotating 0.5 turns between the ends of the girdle, finally broadening and running to the antapex. Flagellar pores at the junctions of the girdle and sulcus. Cytoplasm clear and ± colourless with small refractive bodies and sometimes large food inclusions. Pusules may be present.

Size: 42–75 μm long.

Distribution: Outside Plymouth Sound, frequent in summer; also from the Pacific off California.

GYMNODINIUM Stein

Stein, 1878, p. 95.

The name *Gymnodinium* was first used by Stein for two of Ehrenberg's *Peridinium* species, *P. fuscum* and *P. pulvisculus*. In 1883 Stein transferred the latter to *Glenodinium*, described two new *Gymnodinium* species, and transferred another into the genus. By the time Kofoid & Swezy published their monograph on the 'Unarmoured Dinoflagellata' (1921), 76 species of *Gymnodinium* were known. Now, over two hundred freshwater, brackish and marine forms have been ascribed to the genus. Together with other genera of naked dinoflagellates this genus contains species which are incompletely described as a result of their sensitivity, which causes them to alter shape when examined under the microscope, and fixation which also produces changes in morphology. Culture studies have shown that wide variations in shape can occur and a clone with individuals exhibiting *Gymnodinium* and *Gyrodinium* characters was described by Kimball & Wood, (1965).

Gymnodinium species lack a thickened theca, although some species with such a feature have been included erroneously in the genus; they have a more or less centrally positioned girdle which is not displaced more than one fifth total body length. They may or may not have chloroplasts. The sulcus may extend from the apex to the antapex or may be restricted to the hypocone; it is usually without torsion. The nucleus is usually centrally positioned. The cell surface may be striate, smooth, or punctate. Cyst formation occurs and the cyst generally consists of a thin, hyaline membrane. Kofoid & Swezy (1921) divided the genus into three subgenera, *Gymnodinium, Lineadinium* and *Pachydinium* based on surface markings. They also noted that some cells ascribed to the genus may be free, non-thecate stages of Peridiniaceae. Lebour (1925) also grouped the species together on the basis of their surface markings and Schiller (1933) also followed Kofoid & Swezy's classification. It is to be hoped that the taxonomy of this genus, which can form toxic blooms and are frequently present in large numbers, will one day be elucidated.

Type species: *G. fuscum* (Ehrenberg) Stein.

KEY TO THE SPECIES OF GYMNODINIUM

1a	Chloroplasts absent, cell colourless or pale greenish colour	2
b	Chloroplasts present	8
2a	Surface of cell with longitudinal striations	3
b	Surface of cell apparently smooth	5
3a	Nucleus situated in centre of cell, epicone rounded *G. heterostriatum*	
b	Nucleus situated in hypocone	4

4a	Cell colourless, sulcus only just enters epicone	G. achromaticum
b	Cell colourless, epicone considerably shorter than hypocone	G. filum
c	Cell bluish-pink, sulcus runs almost to apex	G. abbreviatum
5a	Cell rounded, epicone larger than hypocone	G. minus
b	Cell dorso-ventrally flattened	6
6a	Cell with tri-lobed appearance	G. marinum
b	Cell bilobed or ovoid	7
7a	Sulcus extending from apex to antapex	G. variabile
b	Sulcus restricted to hypocone	G. incertum
8a	Cell large, over 50 μm long	9
b	Cell of medium size, 20–40 μm long	G. variabile
c	Cell small, usually under 20 μm long	10
9a	Chloroplasts radiating; girdle ± median	G. splendens
b	Chloroplasts small, scattered; girdle towards anterior end of cell	G. conicum
10a	Cell very small, usually under 10 μm; sulcus not obvious	G. simplex
b	Cell slightly larger, girdle and sulcus obvious	11
11a	Epicone notched	G. pigmaeum
b	Epicone entire and rounded	12
12a	Sulcus cuts epicone asymmetrically	G. micrum
b	Sulcus cuts epicone ± symmetrically	G. vitiligo/G. veneficum

Gymnodinium abbreviatum Kof. & Swezy

Fig. 9A

Kofoid & Swezy, 1921, p. 180, pl. 6, fig. 63, text fig. Z, 7.
Lebour, 1925, p. 48, pl. 6, fig. 6.
Schiller, 1933, p. 324, figs. 328a, b.
Drebes, 1974, p. 118, figs. 96a-c.

The body is long, almost twice the maximum diameter; asymmetrically ovoidal and circular in cross-section. Epicone is much shorter than hypocone and sub-conical with a slightly excentric rounded apex. The hypocone is long, widest anteriorly tapering posteriorly with antapex slightly notched by the sulcus and asymmetrical. The widest part of the cell is at the girdle which is premedian, and displaced 0.26 transdiameters sinistrally. The girdle is not deeply excavated and has overhanging borders. The sulcus runs from the apex to the antapex in an almost straight line. The anterior flagellum pore is at the anterior junction and the longitudinal flagellum pore at the posterior junction of the girdle and sulcus. Pusules may or may not be present at these pores. The nucleus is large and situated in the hypocone. Chromatophores are absent and the cell is a bluish pink in colour. Nutrition holozoic (Lebour figures an individual with an ingested *Cochlodinium*). The cell covering is characteristically thick, giving the outline a wavy appearance and on the surface of the body are longitudinally arranged broken lines of striae.

Size: 85–120 µm long; 50–75 µm diameter.

Distribution: English Channel; elsewhere, Helgoland and List, and La Jolla, California. Probably an oceanic species.

Note: This species closely resembles *G. achromaticum* described from a single specimen by Lebour (1917) but differs in surface sculpturing, colour and in having a sulcus extending from apex to antapex; in Lebour's species it extends from girdle to antapex.

Gymnodinium achromaticum Lebour

Fig 9H

Lebour, 1917, p. 190, fig. 5.
Schiller, 1933, p. 325, fig. 329.

Cell asymmetric, epicone short and conical with a rounded apex, hypocone rather square with antapex cleft by sulcus. Girdle in anterior half of cell, horizontal with slight offset. Sulcus only just extending into epicone, straight. Nucleus near posterior end of cell; cytoplasm colourless. Cell surface striate.

Size: 78 µm long; c 48 µm wide.

Distribution: Plymouth Sound.

Gymnodinium conicum Kof. & Swezy

Fig. 9B

Kofoid & Swezy 1921, p. 198, fig. X27.
Lebour 1925, p. 43, pl. 5, fig. 2.
Schiller 1933, p. 345, fig. 350.

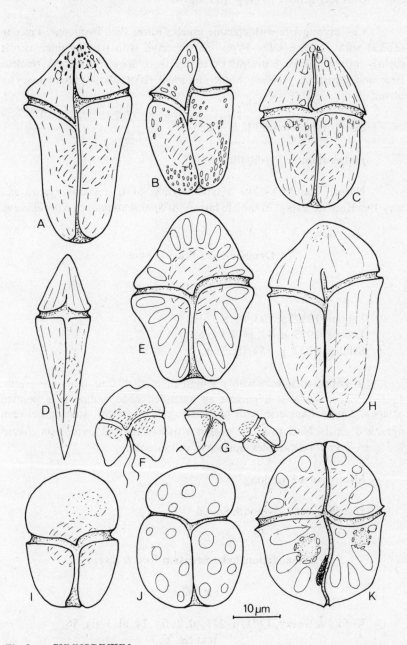

Fig. 9 GYMNODINIUM
A. *G. abbreviatum* B. *G. conicum* C. *G. heterostriatum* D. *G. filum*
E. *G. splendens* F. *G.* stages of division H. *G. achromaticum* I. *G. minus*
J. *G. marinum* K. *G. variabile* (after Dragesco 1965)

Syn: *G. viridis* Lebour 1917, p. 189, fig. 4.

Cell asymmetric with epicone much shorter than hypocone. Epicone conical with concave sides; hypocone trapezoid with straight sides. Girdle slightly offset; sulcus ± straight extending from apex to antapex. Nucleus large and central in position; numerous small chloroplasts present in cytoplasm.

Size: 60 μm long; 32 μm wide.

Distribution: Plymouth Sound.

Note: This looks like a small version of *G. pseudonoctiluca* and may therefore be a stage in the life history of *Spatulodinium* or *Kofoidinium*.

Gymnodinium filum Lebour

Fig. 9D

Lebour, 1917, p. 193, fig. 9.
Lebour, 1925, p. 42, pl. 4, fig. 6.
Schiller, 1933, p. 357, fig. 363.

A slender organism with a length more than four times the width. Epicone short, conical; hypocone an extended cone ending in a pointed antapex. Girdle transverse with no offset; sulcus straight, starting between apex and girdle. Nucleus in the posterior half of the cell; cytoplasm colourless apart from food vacuoles. Surface of cell smooth.

Size: 50–65 μm long.

Distribution: Plymouth Sound.

Gymnodinium heterostriatum Kof. & Swezy

Fig. 9C

Kofoid & Swezy, 1921, p. 221, pl. 2, fig. 24, pl. 5, fig. 56,
                      text fig. Y, 7.
Lebour, 1925, p. 47, pl. 6, fig. 2.
Schiller, 1933, p. 370, figs. 376a-c.
Drebes, 1974, p. 118, figs. 97a-b.

Syn: *G. spirale* var. *obtusum* Dogiel, 1906, p. 38, pl. 2, figs. 50–56.
        ? *G. lucidum* (= *G. hyalinum* Lebour, 1925) Ballantine 1964, p 518.
        ? *G. rhomboides* in Lebour 1925, p. 47, pl. 6, figs. 1a-c.

Cell with rounded or conical epicone and rounded hypocone which are subequal. Epicone usually smaller than hypocone and body almost circular in transverse section. Girdle is submedian, displaced its own width and its ends slightly overlap. The sulcus is narrow and extends from the apex to the antapex becoming very shallow and gradually fading away near the antapex. The anterior pore is at the anterior junction of the girdle and sulcus where a pusule may be present and the posterior pore is situated to the posterior of the junction between the right side of the girdle and the sulcus. The nucleus lies to the anterior of the centre. Chromatophores are absent and the cell contents are colourless or pale yellow. Nutrition is holozoic and cells are frequently found with ingested dinoflagellates. The surface is characteristically covered with longitudinal striations having twice the number on the epicone than on the hypocone. Kofoid & Swezy, and Lebour, found individuals enclosed in thin-walled hyaline cysts.

Size: 40–85 μm long.

Distribution: Plymouth; Naples, La Jolla, California. Probably a warm water form.

Note: According to Drebes and Elbrachter (1976) this species includes *G. lucidum* (= *G. hyalinum* Lebour, 1925) and *G. rhomboides* Lebour. *G. rhomboides* in addition to the locations for *G. heterostriatum* has been recorded from Skagerak, Atlantic, Baltic, from the Outer Hebrides, English Channel and N.E. coast of Britain.

Gymodinium incertum Herdman

Fig. 10D

Herdman, 1924, p. 80, fig. 32
Lebour, 1925, p. 41, fig. 11f.
Schiller, 1933, p. 372 fig. 379.

A small species which is dorso-ventrally flattened. Epicone hemispherical in ventral view, slightly larger than the hypocone. Hypocone short and rounded. Girdle transverse with slight displacement on right side; sulcus short, not extending onto the epicone nor reaching the antapex. Nucleus central, cytoplasm colourless apart from one or two reddish bodies.

Size: 15 μm long.

Distribution: In damp sand at Port Erin (Isle of Man)

Gymnodinium marinum Kent

Fig. 9J

Saville-Kent, 1880/82, p. 444, pl. 25, figs. 60–61.
Kofoid & Swezy, 1921, p. 232, text fig. X, 13
Lebour, 1925, p. 39, fig. 116.
Schiller, 1933, p. 382, fig. 391.

Small species with trilobal appearance in ventral view, compressed dorso-ventrally to show convex dorsal and concave ventral surface. Epicone hemisphaerical with symmetrically rounded sides. The hypocone is slightly longer and wider than the epicone with rounded antapex and rounded sides. The girdle is premedian, deeply impressed and circular. The sulcus is deep and extends from the girdle to the antapex in a straight line. The longitudinal flagellum arrises at the anterior end of the sulcus. Plasma clear and colourless containing numerous 'spherules'. Holozoic nutrition, seen by Saville-Kent to eat *Heteromita* and other monads. Nucleus not figured but thought to be anterior.

Size: 30 μm long, 28 μm wide. Recent specimens assigned to this species 16 μm x 16 μm.

Distribution: From an infusion of hay and sea-water made at St. Helier, Jersey by Saville-Kent. Also believed to have been seen at weather station 'India' in the North Atlantic.

Gymnodinium micrum (Leadb. & Dodge) Loeblich III

Fig. 10A

Syn: *Woloszynskia micra* Leadbeater & Dodge, 1966, p. 1, fig. 1.

Cell broadly oval to round, epicone slightly smaller than hypocone. Girdle deeply incised, displaced into the hypocone on the right ventral side by a projection of the epicone. Sulcus extending a short distance into the epicone and beyond the antapex of the hypocone. Transverse flagellum possessing an expanded sheath, striated accessory strand and unilateral array of long hairs, extending completely around the cell and attached at the distal end in living cells. Longitudinal flagellum 1–1½ times body length, the distal third narrow with bilateral array of short hairs. Cell surrounded by flexible theca composed of small hexagonal units. Nucleus subspherical, of typical dinoflagellate type, 4-6 μ diameter, median to posterior in position. Chloroplasts yellow-green, cup-shaped and much lobed, lamellae usually consisting of 3 thylakoids, several pyrenoids between the lamellae. Associated with the flagella bases are two pusules. Many trichocysts are found in the periphery of the cell. Asexual fission takes place in the motile state.

Size: 9–15 μm long; 8–14 μm wide.

Distribution: English Channel off Plymouth; probably fairly common but normally missed due to its small size and difficulty of preservation.

Note: A detailed fine structural description has been given for this species (Leadbeater & Dodge, 1966).

Gymnodinium minus Lebour

Fig. 9I

Lebour, 1917, p. 192, fig. 8
Lebour, 1925, p. 38, pl. 4, fig. 3.
Schiller, 1933, p. 383, fig. 393.

Cell roundish with the widest part towards the anterior. Epicone slightly larger than hypocone, rounded; hypocone narrower with a truncated antapex. Girdle horizontal, slightly displaced on right side. Sulcus straight, running only from girdle to antapex. Nucleus central, cytoplasm colourless apart from food vacuoles.

Size: 28 μm long

Distribution: Plymouth Sound

Gymnodinium pygmaeum Lebour

Fig. 10F

Lebour, 1925, p. 38, pl. 4, fig. 4.
Schiller, 1933, p. 403, fig. 422.

A very small rounded species with an epicone which is shorter than the hypocone. Both apex and antapex are rounded and notched by the sulcus; girdle transverse and rather broad. Nucleus situated in the epicone; several greenish chloroplasts present.

Size: 14 μm long

Distribution: English Channel

Note: It would seem possible that this is an early report of what has come to be known as *Gyrodinium aureolum* Hulburt, for this organism can be abundant in the area from which *G. pigmaeum* was collected.

Gymnodinium simplex (Lohmann) Kof. & Swezy

Fig. 10E

Kofoid & Swezy, 1921, p. 256, fig. BB, 8.
Lebour, 1925, p. 37, fig. 10.
Schiller, 1933, p. 413, fig. 433.
Dodge, 1974, p. 54, p. 171, pl. 1–4, text fig. 1.
Parke & Dodge, 1976, p. 544.

Syn: *Protodinium simplex* Lohmann, 1908, p. 264, pl. 17, fig. 17.

Cell broadly ellipsoidal with epicone narrower and shorter than hypocone and circular in cross-section. The wide gently depressed horizontal girdle is slightly anterior to the centre of the cell. The sulcus is very shallow and not always discernible and does not extend onto the epicone. The anterior of the cell is rounded and the hypocone has rounded sides with the antapex also rounded, flattened or indented by the sulcus. The spherical nucleus is centrally or slightly posteriorly positioned. Yellow-green chloroplasts normally present, usually two or four situated at the periphery of the cell, occasionally more or they may be absent. Trichocysts are present.

Size: 7–9 μm long (2–25 μm reported)
6–7 μm wide (1.7–13 μm reported)

Distribution: English Channel, N.E. and N.W. coast of England, N. Atlantic, Baltic, Pacific, off U.S.A., S. America and Australasia in temperate, tropical and arctic waters.

Note: This organism has been redescribed with the aid of electron microscopy (Dodge, 1974).

Gymnodium splendens Lebour

Figs. 9E–G

Lebour, 1925, p. 43, pl. 5, fig. 1.
Schiller, 1933, p. 417, fig. 438.
Biecheler, 1952, p. 28, fig. 7.
Steidinger, 1964, pl. 1–4.
Drebes, 1974, p. 117, figs. 95a-c.
Parke & Dodge, 1976, p. 544.

Cell ovoid, flattened dorso-ventrally with convex dorsal surface and concave ventrally. Epicone and hypocone nearly equal. In ventral view the epicone is in the shape of a blunt, rounded cone and the hypocone is deeply indented by the sulcus. The girdle is impressed, left-handed displaced one to two girdle widths. The sulcus expands posteriorly and according to Lebour

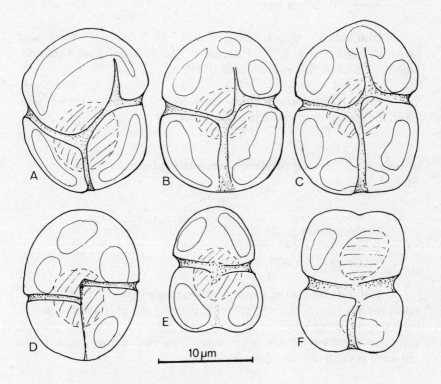

Fig. 10   GYMNODINIUM (note enlarged scale)
A. *G. micrum* B. *G. vitiligo* C. *G. veneficum* D. *G. incertum* E. *G. simplex*
F. *G. pigmaeum*

(1925) and other authors, it does not extend onto the epicone. Biecheler (1952), however, noted a very narrow prolongation of the sulcus onto the epicone. The nucleus is slightly anterior of the midline. Numerous yellow-brown chromatophores radiate from outside the centre to the periphery. Longitudinal flagellum is about one and a quarter times the body length. Reproduction is by longitudinal oblique fusion. Toxic blooms have been recorded by Drebes; it is thought that the organism itself is non-toxic and that shell fish deaths were caused by the *Gymnodinium* clogging the gills (Steidinger, 1964).

Size: 40–80 μm long

Distribution: Coastal form. Found off South and North west coast of England and N.W. of Scotland. Also recorded from California, Florida, Gulf of Mexico and East coast of U.S.A.

Note: This species closely resembles *G. nelsonii* Hulburt from which it differs in the shape of its chloroplasts.

Gymnodinium variabile Herdman

Fig. 9K

Herdman, 1924, p. 80, figs. 35–45.
Lebour, 1925, p. 41, fig. 11e.
Schiller, 1933, p. 424, fig. 446.

Cell rounded in ventral view, dorso-ventrally flattened. Girdle median, sulcus usually short, scarcely entering epicone and only rudimentary or extending to antapex on hypocone. Nucleus spherical, in central position, cytoplasm colourless or pale yellow apart from two or three orange bodies and some greenish or brown granules.

Size: 20–40 μm long.

Distribution: In damp sand at Port Erin, also found on beaches to the south and west of the British Isles and in Brittany.

Note: As originally described this organism is either very variable or a complex of several species. The cell illustrated (Fig. 9K) comes from Dragesco (1965) who maintains that there are both chloroplasts and pyrenoids.

Gymnodinium veneficium Ballantine

Fig. 10C

Ballantine, 1956, p. 467, figs. 6–17.

Cell ovoid and not flattened with equal epicone and hypocone. Epicone with convex sides is slightly pointed at the apex. The hypocone is rounded with a very slight sulcal indentation at the antapex. The girdle is deeply impressed, displaced about one girdle width. The sulcus extends onto the epicone and hypocone and decreases in depth towards apex and antapex. The two flagella are situated one at each junction between the girdle and the sulcus. The nucleus is median. Cytoplasm hyaline, containing refractile oily bodies. Periplast rigid apparently unstructured. Usually four golden brown irregularly shaped chromatophores present, occasionally two to eight and in old cultures colourless cells may occur. Longitudinal flagellum one and a half times cell length. Pusules not observed. Reproduction by fission in the motile state, detailed description given by Ballantine (1956). This species produces an exotoxin which is lethal to fish and other animals.

Size: 9–18 µm x 7–14 µm

Distribution: Isolated from sea water off Devonport (lat. N. 50°21' 50", long W. 04°10'55") by Dr Mary Parke. Type culture at Marine Biological Association, Plymouth (Plymouth collection No. 103) and Cambridge.

Note: This species is very similar to *G. vitiligo* Ballantine (1956), but differs from it in being toxic, having a more pointed epicone, no flap on right ventral epicone, girdle less displaced and a longer flagellum.

Gymnodinium vitiligo Ballantine

Fig. 10B

Ballantine, 1956, p. 468, figs. 1–5.
Parke & Dodge, 1976, p. 544.

Cell slightly longer than broad with rounded apex and antapex, not dorso-ventrally flattened. Epicone slightly smaller than hypocone. The right ventral side of the epicone is produced into a flap which protrudes and partially overlaps the girdle and sulcus. The antapex is not indented by the sulcus. The girdle is deeply impressed, narrower on the ventral side and displaced approximately two girdle widths. The sulcus extends onto epicone and hypocone and gradually decreases in depth distally. The anterior and posterior pores occur at the junctions of the girdle and sulcus. The nucleus is median. Cytoplasm hyaline, containing refractile oily bodies. Periplast rigid apparently unstructured. Usually four golden brown chromatophores of irregular shape present, occasionally two to eight or more or, in old cultures, colourless cells may occur. The longitudinal flagellum is fine and one to one and a quarter times the cell length. The organism swims forward in a jerky irregular spiral. Pusules have not been observed. Reproduction is by fission in the motile state, no encystment or sexual reproduction has been observed. This species is not toxic to fish or other animals.

Size: 7–18 µm x 7–14 µm

Distribution: Isolated by Dr Mary Parke from sea water off Plymouth Sound (lat. N. 50°19'30", long W. 04°10'). Type culture at Marine Biological Association, Plymouth (Plymouth Collection No. 102). Also recorded from N.W. Scotland.

Note: This species very closely resembles *G. veneficium* also described by Ballantine (1956). It differs in not being toxic, having a more rounded apex, more deeply impressed sulcus, more displaced girdle, having a prominent flap on the right ventral epicone and a shorter longitudinal flagellum.

GYRODINIUM Kof. & Swezy

Kofoid & Swezy, 1921, p. 121, p. 273.

Syn: *Gymnodinium* Schutt, 1895, partim
        Spirodinium Schutt, 1896

Description

Girdle characteristically descending in a left spiral displaced more than 0.2 of the total body length. Sulcus longitudinal, or with torsion of less than half a transdiameter in the intercingular region, extending from girdle or epicone to hypocone or antapex. The nucleus is usually situated near the centre. Pusules may be present. Chromatophores rarely present. Surface smooth or with longitudinal striations. Cell shape variable but frequently ovoid or fusiform with girdle centrally positioned although epicone and hypocone can be subequal. Cells vary in length from 11 μm to 200 μm. They occur in marine, brackish or fresh water and are pelagic, neritic or littoral (living in sand).

Two sorts of cysts have been described, a thin-walled capsule in which the individual cell retains its shape before fission and a sphaerical cyst where the individual cell rounds off, loses its flagella and is resistant (Silva, 1959).

Type species: *G. spirale* (Bergh) Kof. & Swezy.

Problems with identification

This is clearly an artificial genus for when Kofoid & Swezy (1921) set it apart from the genus *Gymnodinium* the main criterion was that the girdle is displaced more than one fifth the length of the body. Consequently, identification of members of this genus can be problematical since the shape frequently changes under the microscope and with fixation. Shape can also be affected by changes in salinity. It is also difficult to separate this genus from *Gymnodinium*. Bursa (1962) has shown that *Gyrodinium californicum* Bursa divides to give two daughter cells, one resembling a *Gyrodinium*, the other a *Gymnodinium*. Kimball & Wood (1965) working with an organism identified as *Gymnodinium mirabile* Penard noted different forms resembling *Gyrodinium fissum* (Levander) Kof. & Swezy, *Gyrodinium resplendens* Hulburt, *Gyrodinium aureolum* Hulburt and *Gymnodinium nelsoni* Martin in culture. *Gyrodinium aureolum* was found to become *Gymnodinium*-like on preservation and when left in continuous dark (Braarud & Heimdal, 1970).

KEY TO THE SPECIES OF GYRODINIUM

1a	Chloroplasts present	2
b	Chloroplasts absent	6

2a	Chloroplasts arranged radially	3
b	Chloroplasts scattered or peripherally arranged	4
3a	Cell oval in outline, nucleus in epicone	*G. uncatenatum*
b	Cell rather angular; chloroplasts evenly distributed; sulcus curved on epicone	*G. resplendens*
c	Cell rounded at anterior end, grooved by sulcus at posterior; nucleus central	*G. aureolum*
4a	Cell large (over 80 μm) and rather knobbly in outline	*G. falcatum*
b	Cell of moderate size	5
5a	Epicone conical	*G. prunus*
b	Epicone rounded	*G. aureolum*
6a	Surface of cell striate	7
b	Surface of cell smooth	19
7a	Rodlets, blue-green bodies, present under cell covering	8
b	Rodlets not present	11
8a	Cell considerably (> x 3) longer than wide	9
b	Cell not more than twice as long as wide	10
9a	Cell tapering evenly to points at both ends	*G. fusiforme*
b	Cell somewhat pear-shaped	*G. lachryma*
10a	Nucleus in anterior end of cell; cell not more than twice as long as wide	*G. fissum*
b	Nucleus central; cell greater than twice as long as wide	*G. obtusum*
11a	Cell greater than three times as long as wide	12
b	Cell less than twice as long as wide	15
12a	Anterior end of cell rounded; striations confined to epicone	*G. cuneatum*
b	Anterior end pointed; striations over all of cell	13

13a	Posterior end rounded and cleft by sulcus	*G. crassum*
b	Posterior end pointed or angular	14
14a	Nucleus in anterior half of cell; cytoplasm often reddish in colour	*G. britannicum*
b	Nucleus in centre of cell; cytoplasm pale yellow or green	*G. spirale*
15a	Girdle entirely in posterior half of cell	*G. glaucum*
b	Girdle more or less equally in both halves of cell	16
16a	Cell twisted with epicone pointing to the (its) left	17
b	Cell straight or with epicone pointed to right	18
17a	Cell pear-shaped	*G. pepo*
b	Cell ± ovoid with slight apical protrusion	*G. opimum*
18a	Nucleus in anterior end of cell	*G. pingue*
b	Nucleus in posterior end of cell	*G. cochlea*
19a	Cell small (< 20 μm) rounded at both ends	*G. lebourae*
b	Cell larger; pointed or conical at one end	20
20a	Posterior end flattened with sulcal depression	*G. linguliferum*
b	Posterior end conical	*G. pellucidum*

Gyrodinium aureolum Hulburt

Figs. 11B–D

Hulburt, 1957, p. 209, pl. 2, figs. 8, 9.
Braarud & Heimdal, 1966, p. 93, fig. 2.
Ballantine & Smith, 1973, p. 235, figs. 1–3.
Drebes, 1974, p. 121, pl. 100.

Cell roundish-oval, slightly variable in shape, somewhat flattened dorso-ventrally. In dorso-ventral view the epicone is rounded in surface outline, conical or truncate. The hypocone has convex sides, is flattened

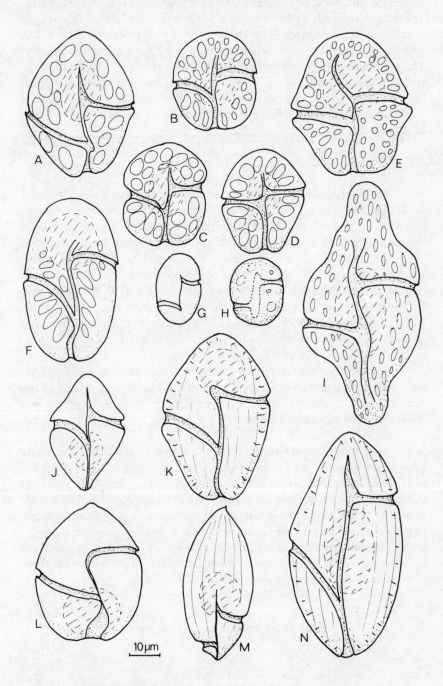

Fig. 11 GYRODINIUM
A. *G. prunus* (after Drebes) B. *G. aureolum* – (after Hulburt 1957)
C. *G. aureolum* – (after Ballantine & Smith 1973) D. *G. aureolum* – (after Drebes 1974) E. *G. resplendens* (after Hulburt) F. *G. uncatenatum* (after Hulburt) G. *G. lebourae* – (after Herdman 1924) H. *G. lebourae* – (after Lee 1977) I. *G. falcatum* J. *G. pellucidum* K. *G. fissum* L. *G. linguliferum* M. *G. glaucum* N. *G. obtusum*

antapically and has a slight indentation where the sulcus reaches the antapex. The epicone is usually slightly smaller than the hypocone. The girdle is deeply incised, wide, left-handed (displacement 0.20 body length). The sulcus extends from half to three-quarters of the way up the epicone to the antapex. The longitudinal flagellum is long, up to twice the body length. Chromatophores are yellowish-brown, elliptical or irregular, numerous and distributed throughout the cell, sometimes in a radiating fashion. The nucleus is large and situated in the girdle region extending into the epicone and hypocone.

Size: 24–40 μm long x 17–32 μm wide.

Distribution: Found in large numbers off the west and south coasts of Britain and west of Ireland, particularly in late summer/early autumn; also off Flamborough and Aberdeen in the east. It would appear to flourish particularly well in frontal systems at the edge of areas of stratification. These are now the subject of detailed study (Holligan & Harbour, 1977; Pingree *et al.*, 1976, 1977). Also recorded from Woods Hole where it was first described, Helgoland, and off the Norwegian coast. In the Oslo Fjord its abundance was associated with the death of *Salmo trutta*, sea trout (Braarud & Heimdal, 1970). Tangen (1977) has carried out a detailed survey of the occurrence of this organism in N. European waters.

It is striking that this species which is now found in such large numbers, went for so long unrecorded. Could it be that it has recently arrived in the area or is it simply that with the coarse net-haul methods of the past it was either not collected or not preserved?

Note: The above description refers to cells in the living state. When being observed under the microscope cells tend to round off. This effect has been noted when cells are shaken or centrifuged; they lose their motility, the characteristic outline and girdle and sulcus almost disappearing (Ballantine & Smith, 1973). Fixation can also distort the shape. A gymnodinioid shape can also be induced by five days continuous dark (Braarud & Heimdal, 1970). Kimball & Wood (1965) consider this species to be synonymous with *Gymnodinium mirabile* or *Gymnodinium fissum* since they have what these authors call a *'Gymnodinium aureolum'* phase in culture.

Gyrodinium britannicum Kof. & Swezy

Fig. 12F

Kofoid & Swezy, 1921, p. 275, 287, fig. DD, 13.
Lebour, 1925, p. 56, pl. 7, fig. 6.
Schiller, 1933, p. 447, fig. 476.
Drebes, 1974, p. 119, figs. 98a-b. (as *G. britanni*).

Syn: *Spirodinium spirale* var. *acutum* Lebour, 1917, p. 194, fig. 10d.

Cell large, spindle-shaped, widest in middle and tapering towards the apices. The hypocone is longer than the epicone which has gently convex sides and a blunt apex. The hypocone has weakly convex sides and ends in a point at the antapex. The girdle starts approximately three-quarters of the way up the body and spirals to the left, being displaced by one or one and a half transdiameters. The girdle is wide and deeply incised. The sulcus extends from the apex to the antapex and is deflected slightly to the left in the girdle region. The nucleus is situated centrally. Chromatophores are absent but the cell is characterised by a carmine red colouring following the longitudinal striae on the cell wall and concentrated at the apices and girdle region. Lebour (1925) says the cell may sometimes be colourless.

Size: 140–170 μm long 44–56 μm wide

Distribution: Plymouth in August

Note: This species is similar to *Gryodinium spirale* from which it differs in having less girdle displacement, blunter apices, and a red colouration which is absent in *G. spirale*.

Gyrodinium cochlea Lebour

Fig. 12E

Lebour 1925, p. 59, pl. 8, fig. 6.
Schiller 1933, p. 452, fig. 481.

Cell elongated with a rather flattened epicone and a pointed hypocone. Girdle making 1.25 turns with the two ends over half the cell-length apart. Sulcus curved, extending from apex almost to the antapex. Cell covering striate; nucleus in posterior half of cell; cell contents hyaline but colourless rods and fat droplets may be present. Chloroplasts absent.

Size: 55 μm long

Distribution: Found in Plymouth Sound in summer.

Gyrodinium crassum (Pouchet) Kof. & Swezy

Fig. 12H

Kofoid & Swezy, 1921, p. 294, fig. CC21
Lebour, 1925, p. 58, pl. 8, fig. 5.
Schiller, 1933, p. 456, fig. 486.

Syn: *Gymnodinium crassum* Pouchet, 1885, p. 66, pl. 4, fig. 28.
Spirodinium crassum Lemmermann, 1899, p. 359.

A large organism with an ellipsoidal body, being rounded at the posterior end and with a blunt projection at the anterior end. Sulcus almost straight, extending from near the apex to the antapex; girdle offset by just under one body width. Cell covering faintly striate, nucleus in posterior half of cell; cell contents a brownish colour and include numerous globules and granules. Chloroplasts absent.

Size: 75–200 µm long; 40–65 µm wide.

Distribution: English Channel, N. Atlantic, Mediterranean.

Gyrodinium cuneatum Kof. & Swezy

Fig. 12B

Kofoid & Swezy, 1921, p. 297, fig. CC17
Lebour, 1925, p. 59, fig. 14C
Schiller, 1933, p. 458, fig. 488.

Cell large, top-shaped, epicone pyramidal with a slightly notched apex, hypocone drawn out and rounded. Sulcus straight, girdle transverse except for the part on the right ventral portion of the cell where it becomes displaced by about one-third the cell length. Epicone striated and containing orange-yellow globules; nucleus situated in the hypocone. Chloroplasts absent.

Size: 90–100 µm long; c 60 µm wide.

Distribution: Coast of Brittany.

Gyrodinium falcatum Kof. & Swezy

Fig. 11I

Kofoid & Swezy, 1921, p. 299, fig. CC11
Lebour, 1925, p. 51, pl. 7, fig. 1.
Schiller, 1933, p. 460, fig. 490.

A large organism with a rather irregular outline. Epicone slightly larger than hypocone, ending in a rounded projection. Hypocone hemi-spherical with a rounded extension. Girdle ± median, displaced by about one-fifth the length of the cell. Sulcus rather short and wide. Nucleus central in position, cell also containing numerous small chloroplasts and oil globules.

Size: 80–120 µm long; 65 µm wide

Distribution: English Channel off Plymouth.

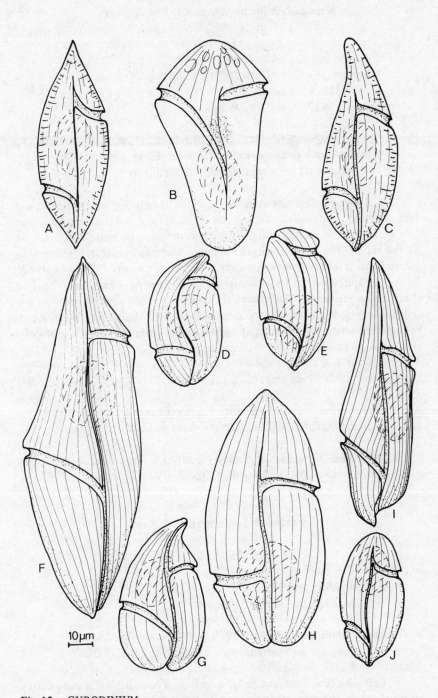

Fig. 12 GYRODINIUM
A. *G. fusiforme* B. *G. cuneatum* C. *G. lachryma* D. *G. opimum*
E. *G. cochlea* F. *G. britannicum* G. *G. pepo* H. *G. crassum* I. *G. spirale*
J. *G. pingue*

10 μm

Gyrodinium fissum (Levander) Kof. & Swezy

Fig. 11K

Kofoid & Swezy, 1921, p. 300, pl. 9, fig. 95, DD 8.
Lebour, 1925, p. 54, fig. 146.
Schiller, 1933, p. 462, fig. 492.

Syn: *Gymnodinium-spirale* var. *D*. Pouchet, 1883, p. 448–449, fig. K.
      Gymnodinium fissum Levander, 1894, p. 43, pl. 2, figs. 5–20.
      Spirodinium fissum Lemmermann, 1900, p. 116.

Cell sub-ovoidal not dorsoventrally flattened. The epicone is longer
than the hypocone with weakly convex sides and a rounded apex. The
hypocone also has weakly convex sides and the rounded antapex is indented
by the sulcus. The girdle is displaced about 0.3 transdiameters, is narrow but
deeply incised. The sulcus begins near the right side of the apex and extends
to the antapex in a gently sinuous line. The sphaerical nucleus is situated
centrally or anteriorly. The plasma is coarsely granular with greenish rodlets
radially arranged. At the periphery there is a series of small radially arranged,
blue-green rodlets placed at right angles to the surface. The cell surface is
striated longitudinally. Colour pale green.

Size: 30–57 µm long; 20–38 µm wide

Distribution: Found in the English Channel, also in California,
Gulf of Finland, Aral Sea, off the Brittany coast and in the Adriatic.

Note: A '*Gyrodinium fissum*' stage was found to occur in cultures
of *Gymnodinium mirabile* by Kimball & Wood, 1965.

Gyrodinium fusiforme Kof. & Swezy

Fig. 12A

Kofoid & Swezy, 1921, p. 307, fig. EE 4, 8.
Schiller, 1933, p. 470, figs, 500a-e.

Syn: *Spirodinium fusus* Meunier, 1910, p. 63, pl. 14, figs. 23–26.
                          Conrad, 1926, p. 86, pl. 1, figs. 21–22.

Cell with long fusiform body, length 3–4 transdiameters and pointed
at both apices. The epicone is slightly smaller than the hypocone, forming a
slender cone with straight to concavo-convex sides. The hypocone is also
slender and conical. The girdle begins about one third of the length from
the apex and descends in a left-handed spiral to about a third of the body
length from the antapex. The girdle is shallow and fairly wide although some
illustrations show it as relatively deeply incised. The sulcus is shallow and

incompletely described. The nucleus is elliptical and is situated centrally or anteriorly. Chromatophores (?) absent. The cell surface is fairly strongly striate and rodlets are present under it all around the cell.

Size: 50–125 μm long; 13–26 μm wide.

Distribution: Found around most of the British Isles. Also recorded from Arctic waters, the Baltic and from Indonesian waters, S. California and the Arabian Sea.

Gyrodinium glaucum (Lebour) Kof. & Swezy

Fig. 11M

Kofoid & Swezy, 1921, p. 308, fig. DD, 16, pl. 9, fig. 94.
Lebour, 1925, p. 54, pl. 7, fig. 4, text fig. 15.

Syn: *Spirodinium glaucum* Lebour 1917, p. 196, figs. 13a-f.
      Massartia glauca Schiller, 1933, p. 436, figs. 462a-e.
                  Hulburt, 1957, p. 208.
                  Sousa e Silva, 1968, p. 33, pl. 7, figs. 2, 3.
      Katodinium glaucum Loeblich III, 1965, p. 15.
                  Drebes, 1974, p. 122, figs. 102a-b.

Cell spindle-shaped, tapering at both apices. Length about 2.5 transdiameters, nearly circular in cross-section. Epicone characteristically much longer than hypocone, apex pointed, sometimes slightly twisted. Hypocone very short, conical and pointed with antapex notched by sulcus. Girdle broad, deeply incised and displaced three to four girdle widths. It commences 0.6 and ends 0.9 of the total body length from the apex. The sulcus runs from the epicone just above the anterior end of the girdle to the antapex, where it broadens. The flagella pores occur at the two junctions of the sulcus and girdle. The elliptical nucleus is situated to the posterior of the centre. Chromatophores are absent, plasma colourless, with a few greenish rodlets in the epicone and usually a bright yellow body at the extreme anterior end which Lebour (1917) suggests may be a product of food formation. Longitudinal striae are indistinct. Division frequently observed.

Size: 40–56 μm long.

Distribution: North Sea in July and August, Plymouth. Elsewhere recorded from La Jolla, California, the Adriatic and the Woods Hole area.

Note: This species is here placed in the genus *Gyrodinium* in agreement with Kofoid & Swezy (1924) and accepted by Lebour (1925). Although the hypocone is very much smaller than the epicone the spiral nature of the girdle shows this to have its affinity with *Gyrodinium* rather than *Katodinium* where it was placed by Loeblich (1965).

Gyrodinium lachryma (Meunier) Kof. & Swezy

Fig. 12C

Kofoid & Swezy, 1921, p. 314, fig. EE, 6.
Lebour, 1925, p. 60, fig. 14e.
Schiller, 1933, p. 475, figs. 505a-c.
Drebes, 1974, p. 121, fig. 101c.
Balech, 1976, p. 21, fig. 12.

Syn: *Spirodinium lachryma* Meunier, 1910, pl. 14, figs. 21, 22.

Cell ovoid, broadly rounded posteriorly tapering to a point anteriorly. Length 2.83 transdiameters. The epicone is longer than the hypocone but narrower in width. The sides when seen in dorsoventral view are convex near the girdle, becoming concave towards the apex. The hypocone is hemisphaerical. The girdle starts about the middle of the body length and forms a left-handed spiral displaced 1.4 transdiameters. It is incised and narrow. The sulcus runs from the anterior end of the girdle to just above the antapex. The nucleus is large, ellipsoidal and positioned in the centre of the body with its long axis parallel to the longitudinal axis of the cell. Chromatophores absent.

Size: 80–105 µm long

Distribution: Occurs in the North Sea from March to November. Also recorded in Arctic waters from the Kara Sea; from Antarctic waters and off the eastern coast of the U S A.

Gyrodinium lebouriae Herdman

Fig. 11G, H

Herdman, 1924, p. 81, fig. 28
Lebour, 1925, p. 53, fig. 14f.
Schiller, 1933, p. 476, fig. 506.
Lee, 1977, p. 303, figs. 1, 3, pl. 1–4.

Body elliptical or irregularly ovate, flattened dorsoventrally. Epicone rounded, hypocone asymmetrical at antapex where it is indented by the sulcus. Girdle left-handed, displaced approximately one third of the total body length, incised. The sulcus commences on the epicone just above the girdle and reaches the antapex. A process known as the peduncle (Lee, 1977) conceals the sulcus in the central region. The nucleus is situated in the hypocone. Herdman describes the presence of a small red body lying near the posterior end of the girdle. The cell is colourless.

Size: 8–18 µm long   6–11 µm wide.

Distribution: Occurs in sand at port Erin (Isle of Man) and found by Conrad and Kufferath in Station S (Belgium). Also isolated from the Sea off South Africa.

Note: See Lee (1977) for ultrastructural description.

Gyrodinium linguliferum Lebour

Fig. 11 L

Lebour, 1925, p. 25, pl. 7, fig. 2.
Schiller, 1933, p. 477, fig. 507.

A rather small species with a rounded body. Apex rounded-conical, antapex broadly rounded. Girdle submedian offset about one-third of cell length. Sulcus forming a wide furrow on the hypocone but not extending onto the epicone. Cell contents colourless apart from a food mass; nucleus situated in the hypocone.

Size: c 40 µm long

Distribution: English Channel off Plymouth

Gyrodinium obtusum (Schutt) Kof. & Swezy

Fig. 11N

Kofoid & Swezy, 1921, p. 319, fig. DD, 3, pl. 9, fig. 103.
Lebour, 1925, p. 57, pl. 8, fig. 3.
Schiller, 1933, p. 482, figs. 153a, b.

Syn: *Gymnodinium spirale* Bergh var. *obtusa* Schutt, 1895, pl. 22, fig. 70.
Spirodinium spirale var. *obtusum* Lemmerman, 1899, p. 359.

Cell ellipsoidal widest in the middle. Epicone, like hypocone, broadly rounded. Cell length 1.75 transdiameters. Girdle descends in a steep left spiral displaced about one transdiameter and is deeply incised. It starts 0.22 and ends 0.8 of the total body length from the apex. The sulcus extends from the apex to the antapex in almost a straight line. The transverse flagellum pore is situated at the anterior junction of the girdle and sulcus whereas the longitudinal flagellum pore is about one girdle width below the posterior junction. Pusules may be connected to these pores. The ellipsoidal nucleus is central or postero-central. The cytoplasm is granular and amber in colour. In the periphery there are numerous, short, blue-green rodlets. The surface is finely striate.

Size: 50–70 µm long

Distribution: Plymouth.

Also from adriatic, off coast of California, and the Arabian Sea.

Gyrodinium opimum (Schutt) Lebour

Fig. 12D

Lebour, 1925, p. 57, pl. 8, fig. 2.
Schiller, 1933, p. 484, fig. 515.

Syn: *Gymnodinium opimum* Schutt, 1895, pl. 21, fig. 68b.

Cell an asymmetric ovoid with the apex tilted to the left. Epicone with a blunt end, hypocone rounded. Girdle forming a spiral, displaced about one body width. Sulcus sinuous, deflected to the left on the epicone. Nucleus central, cytoplasm colourless. Cell surface delicately striate.

Size: c 50 µm long.

Distribution: Plymouth Sound; also Mediterranean and Atlantic.

Gyrodinium pellucidum (Wulff) Schiller

Fig. 11J

Schiller, 1933, p. 490, fig. 521

Syn: *Gymnodinium pellucidum* Wulff, 1916, p. 107, pl. 1, fig. 2.

Cell rhomboidal, both epicone and hypocone being rather conical. Girdle offset by about one quarter the cell length; sulcus slightly curved. Cell contents clear apart from granules; nucleus in the hypocone. Surface of cell lightly striate. Chloroplasts absent.

Size: 25–40 µm long.

Distribution: Barents Sea, Paciffic Ocean.

Note: Recently it has been suggested that this organism is in fact the gamete stage of *Polykrikos kofoidii* (Morey-Gaines & Ruse, 1980). If this is true it may well be that other *Gyrodinium* species are stages in the life history of other dinoflagellates.

Gyrodinium pepo (Schutt) Kof. & Swezy

Fig. 12G

Kofoid & Swezy, 1921, p. 326, fig. DD 2.
Lebour, 1925, p. 55, pl. 7, fig. 5 *(G. bepo)*,
Schiller, 1933, p. 490, fig. 552.

Syn: *Gymnodinium spirale* var. *pepo* Schutt, 1895, p. 112, fig. 69.
       Spirodinium spirale var. *pepo* Lemmermann, 1899, p. 359.

Body flask-shaped with epicone either straight or curved to the left. Hypocone rounded but grooved by the sulcus. Girdle displaced about one-third of body length; sulcus sinuous. Nucleus in anterior half of cell; cytoplasm colourless. Chloroplasts absent.

Size: 56–84 µm long; c 50 µm wide.

Distribution: Plymouth Sound, also found in the Mediterranean and the Atlantic.

Gyrodinium pingue (Schutt) Kof. & Swezy

Fig. 12J

Kofoid & Swezy, 1921, p. 327, fig. DD, 15, pl. 4, fig. 38.
Lebour, 1925, p. 58, pl. 8, fig. 4.
Schiller, 1933, p. 491, figs. 523a–h.

Syn: *Gymnodinium spirale* var. *pinguis* Schutt, 1895, pl. 21, fig. 65.
       Spirodinium spirale var. *pingue*, Lemmermann, 1899, p. 359.

Cell elongated ovoidal, circular in cross-section, nearly twice as long as broad. The epicone is smaller than the hypocone, convex conical with a blunt apex. The hypocone is elongate with a broad antapex. The girdle is displaced 0.74 transdiameters, commences 0.25 and ends 0.66 of the total body length from the apex. It is deeply incised. The sulcus extends from just below the apex to the antapex in a gently sinuous line, it is narrow, and shallow, fading out near the antapex. The nucleus is sphaerical and situated near the centre of the cell. A large pusule opens into the posterior flagellum pore and the transverse flagellum may also have a pusule. The cytoplasm is dense, containing refractive granules, blue-green oil droplets and food vacuoles. A peripheral layer of blue-green rodlets may be present. The surface of the cell is finely striated.

Size: 45–60 µm.

Distribution: Found off Plymouth.

Also recorded from List, the Atlantic, Mediterranean, Arabian Sea, from the Barents Sea and off the Californian coast.

Gyrodinium prunus (Wulff) Lebour

Fig. 11A

Lebour, 1925, p. 52, fig. 14a.
Schiller, 1933, p. 494, fig. 525.
Drebes, 1974, p. 121, figs. 101a, b.

Syn: *Spirodinium prunus* Wulff, 1920, p. 107, fig. 3.

Cell plum-shaped with epicone slightly smaller than hypocone. The epicone is hemisphaerical and the hypocone more pointed and indented at the antapex by the sulcus. The girdle is deep, narrow and displaced 0.75 transdiameters. It commences at about or a little to the anterior of the centre of the cell and ends about three quarters of the way down the cell, having spiralled to the left. The sulcus commences to the anterior of the girdle on the epicone and runs straight to the antapex. The nucleus is central to anterior. There are numerous yellow-brown chromatophores and vacuoles and refractive bodies may be present. Drebes (1974) described individuals being encased in a jelly which he suggests may be connected with a division stage.

Size: 40–55 µm long.

Distribution: Occurs in spring off Helgoland.

Also recorded from the Adriatic in winter and spring, from the Barents Sea, Straits of Florida and north coast of Brazil.

Gyrodinium resplendens Hulburt

Fig. 11E

Hulburt, 1957, p. 210, figs. 6–7.

Cell broadly ovate with rounded apex and rounded to-slightly-squared antapex. Epicone dome-shaped, hypocone cleft by the widening sulcus. Girdle median, displaced up to a quarter of body length, sulcus extending onto epicone as a very narrow groove twisted to the right. Nucleus just in anterior half of cell; numerous brown chloroplasts present, often radially arranged. Longitudinal flagellum about one cell length.

Size: 36–62 µm long; 32–48 µm wide.

Distribution: First described from Great Pond, Woods Hole (USA). Also found in Swan Pool, Falmouth, Cornwall.

Note: See Hulburt (1957) for more detailed description. This species is said to have some similarities with *Gyrodinium aureolum* and *Gymnodinium nelsonii.*

Gyrodinium spirale (Bergh) Kof. & Swezy

Fig. 12I

Kofoid & Swezy, 1921, p. 332, pl 4, fig. 43, fig. DD, 14.
Lebour, 1925, p. 56, pl. 8, fig. 1.
Schiller, 1933, p. 498, figs. 530a–e.

Hulburt, 1957, p. 213, fig. 4.
Drebes, 1974, p. 119, fig. 98c.

Syn: *Gymnodinium spirale* Berg, 1881, p. 66.
Spirodinium spirale Entz, 1884, p. 39.

This widely distributed species is the type for the genus. The cell is slender, fusiform, widest posteriorly and tapering towards the anterior. Length of body 2.5 transdiameters; body convex dorsally and subconcave ventrally giving an asymmetrical appearance. The cone-shaped epicone is convex on the left and dorsal sides but concave on the right and ventral side. The apex is usually blunt but may be pointed. The hypocone is straight-sided anteriorly, convex posteriorly and deeply notched on the left side of the antapex by the sulcus with the right side forming a rounded lobe which projects beyond the posterior end of the left side. The girdle starts anterior to the midline and steeply spirals to the left with a displacement of about 1.2 transdiameters ending a quarter of the total cell length from the antapex. The girdle is deeply incised and approximately 4–6 µm wide. The transverse flagellum pore is situated at the anterior junction of the girdle and sulcus, the longitudinal flagellum pore about one girdle width below. The sulcus extends from the apex to the antapex, is narrow anteriorly, widening slightly towards the antapex. The nucleus is ellipsoidal to sphaerical, situated in the intercingulum area or to the anterior. Pusules may be present at either pore. Chromatophores absent, plasma colourless, or pale green or pale yellowish brown. The surface is marked with longitudinal striae.

Size: 40–200 µm long (around Britain cells usually 50–103 µm).

Distribution: Port Erin (Isle of Man), Plymouth, Morecambe Bay. Worldwide distribution, Mediterranean, Atlantic, Pacific, Arabian Sea.

Gyrodinium uncatenatum Hulburt

Fig. 11F

Hulburt, 1957, p. 210, figs. 1–3.

Cell flattened laterally, elliptical in lateral view. Epicone helmet-shaped, taller than wide, broadly rounded; hypocone similar but in ventral view grooved by the sulcus. Girdle median, offset by about one-quarter body length; sulcus sinuous only just entering epicone, deeply excavated in hypocone. Posterior flagellum up to twice body length. Nucleus situated in epicone; numerous yellow-brown chloroplasts radiate from centre of cell; many dark bodies also present.

Size: 40–54 µm long; 28–33 µm wide

Distribution: Originally described from Woods Hole, USA. Also found in the Hebrides.

HEMIDINIUM Stein

Stein, 1883, p. 91, 97, pl. 2, figs. 23–26.

Body asymmetrically elliptical, girdle slightly left-handed but only encircling half of the cell. Sulcus (when visible) broadening towards antapex. Cell covered with delicate theca consisting of 17 plates.

Note: A number of species have been removed from this genus and placed in *Bernardinium* (Javornicky, 1962) on the grounds that these have only a structureless periplast and no thecal plates.

Hemidinium nasutum Stein

Fig. 13D

Stein, 1883, p. 91, pl. 2, figs. 23–26.
Lebour, 1925, p. 20, fig. 7.
Woloszynska, 1925, p. 7, figs. 4–5.
Schiller, 1937, p. 89, fig. 75.

Cell asymmetrically ellipsoidal, dorso-ventrally flattened, rounded at anterior end, rounded to slightly depressed at posterior. Girdle ± median around only the left half of the cell, descending to the rear. Sulcus in hypocone, widening towards antapex. Theca containing c. 17 very delicate plates variable in arrangement. Cell contains numerous small yellow-brown chloroplasts. Longitudinal flagellum 1½–2 times body length.

Size: 24–28 μm long; 15–22 μm wide.

Distribution: The genus is essentially one of freshwater but this species is also known from salt lakes and brackish coastal localities.

PTYCHODISCUS Stein

Stein, 1883, p. 28.

Cell compressed apically – antapically. Epicone and hypocone more or less equal in size separated by a wide girdle. Cell covered by a thick periplast not apparently consisting of separate plates.

Note: Recently, Steidinger (1979, p. 440–1) has moved the Florida red tide organism *Gymnodinium breve* into this genus. However, this would seem to be unjustified as electron microscopy (Steidinger & Truby, 1978) shows *G. breve* to have a similar theca to that of other naked dinoflagellates which have been studied.

Ptychodiscus noctiluca Stein

Fig. 13B, C

Stein, 1883, p. 23, pl. 23, figs. 7–10.
Schiller, 1937, p. 76, fig. 60.

Syn: *P. carinatus* Kofoid, 1907, p. 168, pl. 1, figs. 8—9.
      P. inflatus Pavillard, 1916, p. 12, pl. 1, fig. 3.
      Diplocystis antarctica Cleve, 1900.

Body compressed apically — antapically. Epicone flattened with a prominent heel on the ventral side, hypocone domed. Girdle horizontal, wide, deep and bounded by substantial ridges or flaps. Sulcus starting on the heel and forming a groove in the hypocone. Cell contains numerous granules but no chloroplasts. The covering is tough and resists fixation but does not appear to be composed of thecal plates.

Size: about 30 μm long; diameter of hypo- and epicones 50—90 μm.

Distribution: Often found in the plankton associated with salps (e.g. *Pegea* sp.). Found in the western approaches, Bay of Biscay, and to the W. and N. of Scotland. Also recorded from the Atlantic, Pacific and Mediterranean.

Note: This organism has recently been studied by Boalch (1969) who found a considerable variation in the form of a cell in fixed or preserved material. He concluded that the various other species which had been described are fixation artifacts.

TORODINIUM Kof. & Swezy

Kofoid & Swezy, 1921, p. 389.

Syn: *Gymnodinium* Pouchet, 1885, partim

Body elongate, about 4 times as long as broad, epicone several times longer than the hypocone. Girdle forming a slightly displaced left-handed spiral. Sulcus is curved and runs along most of the epicone, not observed on the hypocone. Nucleus said to be elongated. Long pigmented bodies (chloroplasts or rhabdosomes) in the epicone, along the sides and collected in a mass at the apex. A long narrow pusule is associated with the flagellar pore of the longitudinal flagellum

Type species: *T. teredo* (Pouchet) Kofoid & Swezy.

Torodinium robustum Kof. & Swezy

Fig. 13A

Kofoid & Swezy, 1921, p. 391, figs. II, 1—3, pl. 4, fig. 49
Lebour, 1925, p. 66, pl. 10, fig. 1.
Schiller, 1933, p. 545, fig. 575.

Syn: *Gymnodinium teredo* Schutt, 1895, pl. 23, fig. 74 partim
       G. teredo Paulsen, 1908, p. 97 partim

Body elongated fusiform, around 3 times as long as broad. Epicone very long, four-fifths of total length, hypocone reduced to a conical structure. Girdle forming a left-handed spiral. Sulcus commencing near cell apex with a loop nearly completely around the cell then curves to meet the girdle. It has not been observed on the hypocone. The flagella arise from the junction of girdle and sulcus and the longitudinal flagellum is very short.

The nucleus is elongated and situated along the centre of the cell. Also there is an elongated pusule which extends anteriorly from the flagellar pores, this appears pink in live cells. The cell contains numerous refractile granules and a number of elongated yellow bodies of unknown function (? chromatophores, rhabdosomes) with more aggregated into a stellate cluster at the anterior end. The cytoplasm has a greenish colouration. The cell is presumably bounded only by thecal membranes as it rounds up before final collapse.

Size: 36–75 μm long; 21–25 μm wide.

Distribution: A planktonic species which has been found at Burnham (Essex) and Plymouth. Also reported from the North Sea and the Pacific.

Family   Pronoctilucaceae

OXYRRHIS Dujardin

Dujardin, 1841, p. 347, pl. 5, fig. 4.

Syn: *Glyphodinium* Fresenius, 1865, p. 83, figs. 4-10.

Body rather oval with rounded anterior and asymmetric posterior which projects to the left. No clear sulcus or girdle but flagella emerge from either side of a tentacular lobe. Nutrition phagotrophic or saprobic; there are no chloroplasts. Reproduction is by transverse binary fission (unlike most other dinoflagellates which divide longitudinally or obliquely). Body and flagella clothed with small scales (Clarke & Pennick, 1972, 1976). The ultrastructure has been described in detail for *O. marina* (Dodge & Crawford, 1971, 1972, 1974).

Type species: *Oxyrrhis marina* Dujardin

Oxyrrhis marina Dujardin

Fig. 13E, F

Dujardin, 1841, p. 347, fig. 4.

Syn: *O. tentaculifera* Conrad, 1939, p. 4, figs. 4–7.
       O. maritima Van Meel, 1969, p. 5, pl. 2, figs. K, L.

Cell an elongated oval with an asymmetric posterior end and broadly conical anterior end. Cell proportions very variable depending on feeding state as there are no thecal plates. Posterior end of cell projects in the form of a relatively stiff lobe and there is a depression from the centre of which arises a tentacular lobe. The flagella arise either side of this. Both flagella are rather similar in appearance and for part of their length contain a paraxial rod. They bear fine hairs and scales. The surface of the body is also covered by tiny scales which can only be seen by electron microscopy.The nucleus is situated near the anterior end of the cell. Food vacuoles frequently present in the cytoplasm. The cell is colourless or slightly pinkish and nutrition is holozoic. Division by transverse cytokinesis.

Size: 10–35 μm long; width 8–30 μm

Distribution: Having an almost universal distribution in salt-marsh pools, littoral pools and occasionally on sandy beaches. Found all around the British Isles.

Note: This is an extremely variable species with size and shape very dependent on food supply and time since division. On these grounds the two species *O. tentaculifera* and *O. maritima* have been merged with the original species. For its food supply *O. marina* is exceedingly omnivorous and has been found to consume other dinoflagellates cryptomonads, green algae, diatoms, yeasts as well as to exist on a defined medium.

PRONOCTILUCA Fabre-Domergue

Fabre-Domergue, 1889, p. 356, pl. 3, figs. 9–10.

Non-photosynthetic organism with short apical tentacle and two anteriorly placed long flagella. Flexible cell wall present with delicate cyst walls produced on the exterior. There is some dispute over the validiy of the genus as a dinoflagellate (Kofoid & Swezy, 1921; Drebes & Elbrachter, 1976).

Type species: *P. pelagica* Fabre-Domergue.

Pronoctiluca pelagica Fabre-Domergue

Fig. 13G

Fabre-Domergue, 1889, p. 356, pl. 3, figs. 9–10.
Pavillard, 1922, p. 365.
Lebour, 1925, p. 18, fig 6.
Schiller, 1933, p. 268, figs. 258a–i.
Wood, 1954, p. 216, fig. 74.
Taylor, 1976, p. 187, pl. 37, figs. 426–428, 430–431.

Syn: *Pelagorhynchus marina* Pavillard, 1917, p. 238, figs. 1–9.
       Rhynchomonas marina Lohmann, 1902, p. 48, pl. 2, figs. 42–45.
                              Ostenfeld & Paulsen, 1904, p. 26, p. 139.
                              Gran, 1915, p. 1, p. 142.
       Protodinifera marinum Kofoid & Swezy, 1921, p. 115.

Cell fusiform to pyriform, length about two and a half times diameter with a short, slender tentacle at apex. Two long flagella, one directed anteriorly, the other posteriorly, emerge near the anterior of the cell on the ventral surface. There is a short ventral, anterior sulcus. Delicate cyst walls are produced exterior to the cell wall. The nucleus is large and is in part responsible for this species inclusion in the Dinoflagellata. This has recently been questioned by Drebes and Elbrachter (1976) who feel the organism belongs to the Chrysophyta. A large accumulation body can sometimes be seen at the posterior end of the cell.

Size: length 12-45 μm.

Distribution: Found at Port Erin, Isle of Man in May. Also known from the coast of Brittany, Gulf of Lions, Bay of Kiel, Indian Ocean.

Family   Warnowiaceae

NEMATODINIUM Kof. & Swezy

Kofoid & Swezy, 1921, p. 415, 421.

Cell with spiral girdle and sulcus rather like *Gyrodinium* but distinguished from that genus by the presence of a large ocellus (elaborate eye-spot) at the posterior end and a number of nematocysts.

Type species: *N. partitum* Kofoid & Swezy.

Nematodinium armatum (Dogiel) Kof. & Swezy

Fig. 13I

Kofoid & Swezy, 1921, p. 422, figs. NN 1–2.

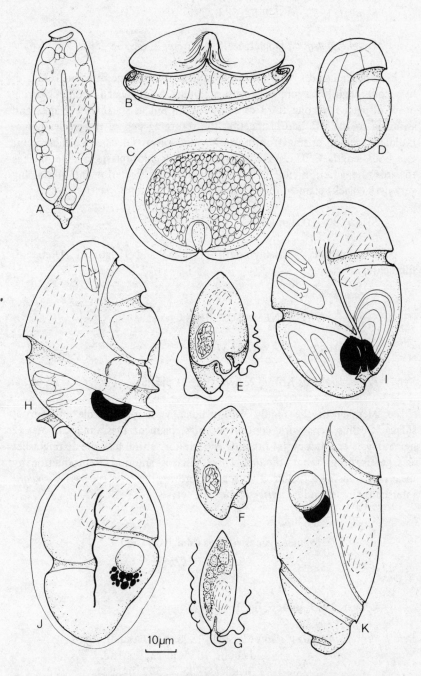

Fig. 13   A. *Torodinium robustum*   B. *Ptychodiscus noctiluca* (after Boalch)
C. *Ptychodiscus noctiluca* apical view   D. *Hemidinium nasutum*   E. F. *Oxyrrhis marina*   G. *Pronoctiluca pelagica*   H. *Nematopsides vigilans* (after Greuet 1972)
I. *Nematodinium armatum*   J. *Protopsis simplex*   K. *Warnowia polyphemus*

Lebour, 1925, p. 71, pl. 10, fig. 5.
Schiller, 1933, p. 559, fig. 588.

Syn: *Pouchetia armata* Dogiel, 1906, p. 36, pl. 2, figs. 38–39.

Body elliptical with ± equal epicone and hypocone. Epicone rounded, hypocone slightly asymmetric with a sloping or depressed antapex. Girdle deep, displaced by one third of body length. Sulcus starting near apex and swinging to the left before meeting the returning end of the girdle. Longitudinal flagellum arising from posterior junction of sulcus and girdle, about one body-length. Cell contains scattered yellow chloroplasts, a large nucleus towards the anterior end, a large ocellus at the posterior end containing very dark (black) pigmentation, and a number of nematocysts.

Size: 30–100 μm long.

Distribution: Off Plymouth (Lebour) and Helgoland (Drebes & Elbrachter).

NEMATOPSIDES Greuet

Greuet 1973, p. 466.

Syn: *Proterythropsis* Kofoid & Swezy, 1921 partim.

Member of the family Warnowiidae, with spiral girdle and curved sulcus. Ocellus, lens with concentric rings, pigment mass and capsules (? nematocysts or cnidocysts) present. A posterior-ventral tentacle distinguishes this genus from *Nematodinium*. Large nucleus anterior. Reproduction by means of transverse fission. Two species known; the type species *N. vigilans* (Marshall) Greuet and *N. tentaculoides* Greuet.

Nematopsides vigilans (Marshall) Greuet

Fig. 13H

Greuet, 1973, p. 466, fig. 10.

Syn: *Proterythropsis vigilans* Marshall, 1925, p. 177, pl. 9, figs. 1–2.
         Lebour, 1925, p. 73, fig. 18d.
         Schiller, 1933, p. 592, fig. 621.
         Drebes, 1973, p. 124–126, fig. 104.

Small, ovoid cell with a left-handed spiral girdle making one and a half turns continuing onto a posterior ventral tentacle. The curved sulcus makes half a turn. Posterior ocellus, red pigment mass with yellow core and pyriform lens with concentric rings present. Elongated capsules each containing six smaller capsules occur. Cell contents pink. The nucleus is large and anterior. Transverse flagellum short, arising in sulcus well posterior to the junction with the girdle; longitudinal flagellum long. Reproduction by transverse fission.

Size: 38–58 μm long

Distribution: Found originally at Millport and elsewhere in the Clyde area. In the present survey recorded from Liverpool Bay in October. Elsewhere recorded from Helgoland April–June.

PROTOPSIS Kofoid & Swezy

Kofoid & Swezy, 1921, p. 415.

Similar to *Gymnodinium* or *Gyrodinium* but containing a simple or compound ocellus. No posterior tentacle nor cell torsion. The girdle does not make more than one turn round the body.

Type species: *P. nigra* (Pouchet) Kof. & Swezy

Note: It has recently been suggested by Greuet (1972) that cells designated as *Protopsis*, which are often found surrounded by a hyaline cyst wall, are in fact stages in the development of *Warnowia* species. As yet the direct relationship has not been established.

Protopsis simplex Lebour

Fig. 13J

Lebour, 1925, p. 70, pl. 10, fig. 4.
Schiller, 1933, p. 558, fig. 587.

Cell normally enclosed in a loose cyst. Resembles a *Gyrodinium* but contains a simple lens which distinguishes it from *P. nigra* and its black pigment is in several separate globular bodies. Girdle displaced about 0.4 transdiameters, sulcus slightly curved extending onto the epicone but not reaching the apices. Plasma yellow. Large nucleus anterior in position.

Size: Length 74 μm,

Distribution: Plymouth.

WARNOWIA Lindmann

Lindemann, 1928, p. 52.

Syn: *Pouchetia* Schutt 1895, pro parte, p. 168, pl. 26, figs. 92, 94.

Un-armoured dinoflagellates with both girdle and sulcus in helices around the cell, girdle 1.1–2 turns and sulcus 0.25–1.5 turns. Ocellus present with large hyaline lens and red or black pigment mass. Nucleus usually in anterior part of cell. No nematocysts present. Nutrition holozoic. Encystment in a thin walled membrane frequently seen.

Type species: *W. fusus* (Schutt) Lindemann

Note: Recent work by Greuet (1972) has suggested that some organisms which have been designated species of *Protopsis* may in fact be stages in the life-cycle of *Warnowia*. There is also confusion about specific identification in this genus and some workers have simply reported *Warnowia* sp.

Warnowia polyphemus (Pouchet) Schiller

Fig. 13K

Schiller, 1933, p. 576, fig. 606.

Syn: *Pouchetia polyphemus* (Pouchet) Kofoid & Swezy, 1921, p. 453.
               Lebour, 1925, p. 72, pl. 10, fig. 6.

Cell elongated, ellipsoidal with a rather square anterior end. Girdle making two turns and sulcus only 0.5 turns. Ocellus in the anterior half of the cell, consisting of a large laminated lens and simple dark pigment cup. Cytoplasm colourless.

Size: 75–100 µm long; c. 47 µm wide.

Distribution: Atlantic Ocean, English Channel.

Note: Four other species have been reported from the waters around the British Isles:– *W. fusus* (Schutt) Lindemann which is spindle-shaped and has a girdle making almost two complete turns; *W. parva* (Lohmann) Lindemann which is much more rounded and compact with a girdle making only one turn and the ocellar lens separated from the pigment globules; *W. rosea* (Pouchet) Kof. & Swezy which is also compact but has a pigment cup attached to the lens; *W. schuetti* (Kof. & Swezy) Schiller which is asymmetric and has a dispersed ocellus, the lens having five segments and the pigment body is dendritic.

Family   Polykrikaceae

POLYKRIKOS Butschli

Butschli, 1973, p. 673—676.

Syn: *Pheopolykrikos* Chatton, 1933, p. 251, fig. 1.

Organisms with a permanent colonial or coenocytic organisation consisting of 2—8 gymnodinioid individuals (zooids) joined in a chain and each having longitudinal and transverse flagella. The number of nuclei is often half that of the number of zooids. Girdles in each zooid slightly offset to the right side of the cell, sulci straight. Cells may be colourless and hetero-trophic or may contain yellow-brown chloroplasts. Nematocysts may be present.

Type species: *P. schwarzii* Butschli

KEY TO THE SPECIES OF POLYKRIKOS

1. a.   Colony ± cylindrical,   2

    b.   Colony laterally flattened with sulcus
         along a narrow side   *P. lebourae*

2. a.   Chloroplasts absent, half number of nuclei
         as zooids   3

    b.   Chloroplasts present, same number of nuclei
         as zooids   *P. beauchampii*

3. a.   Individual zooids not clearly demarkated
         cell covering ± smooth   *P. schwarzii*

    b.   Zooids clearly distinguished, hypotheca
         longitudinally ridged   *P. kofoidii*

Polykrikos beauchampii (Chatton) comb. nov.

Fig. 14E, F

Syn: *Pheopolykrikos beauchampi* Chatton, 1933, p. 251, fig. 1.

Elongate colony, with rounded anterior end and rounded posterior end with sulcul depression, consisting of four zooids. The girdles are slightly offset and secondary constrictions between the zooids are clear or may be very conspicuous, particularly if the organism has been slightly squashed. The sulci form a central groove along the ventral side of the colony. There are four nuclei and numerous chloroplasts and refractile granules. Nematocysts are present and a pusule is associated with each longitudinal flagellum.

Size: 100–120 μm long; 60–75 μm wide.

Distribution: In semi-permanent salt-marsh pool at Newport, I.O.W.; originally described from brackish water in the south of France.

Note: When originally described by Chatton (1933) this organism was placed in a new genus, its main distinguishing features being the presence of chloroplasts and similar number of nuclei and zooids. Neither seem to justify a separate genus and it is here brought into *Polykrikos*.

Polykrikos kofoidii Chatton

Fig. 14D

Chatton, 1914, p. 161.
Schiller, 1933, p. 549, fig. 578.

Syn: *P. schwarzi* Kofoid, 1907, p. 291, fig. 1.

Colony elongated with rounded anterior and flattened posterior, consisting of 2–8 zooids each of which is clearly distinguished by constriction and by the marked longitudinal ridging of the hypothecae. Colony contains two nuclei and several nematocysts but chloroplasts are absent. Colour greenish to pink.

Size: single zooid 25–45 μm long; colony up to 110 μm.

Distribution: Probably an oceanic species; reported off Plymouth; originally described from the Pacific Ocean.

Note: The encystment of motile colonies of this species to give a spiny cyst similar in shape to the cyst of *P. schwarzii*, but with separate spines, has recently been described (Morey-Gaines & Ruse, 1980). These authors have also suggested that *Gyrodinium pellucidum* (Wulff) Schiller may repesent freed sub-units of this organism and possibly also gametes.

Polykrikos lebourae Herdman

Fig. 14B, C

Herdman, 1924, p. 5, fig. 6.
Lebour, 1925, p. 68, pl. 10, fig. 3.

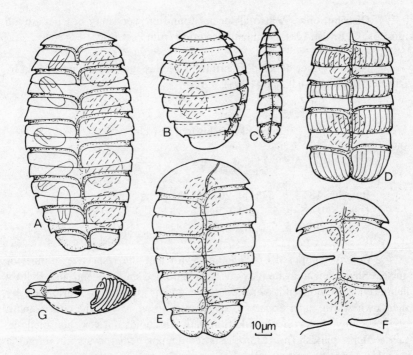

Fig. 14 POLYKRIKOS
A. *P. schwartzii* B. *P. lebourae* side view C. *P. lebourae* ventral view
D. *P. kofoidii* E. *P. beauchampii* F. *P. beauchampii* after squashing
G. Nematocyst from *P. schwartzii* (not to scale) (after Chatton & Greuct)

Syn: *P. schwarzii* Herdman, 1922

Colony broadly ovate in lateral view, widest at the mid point; narrow and twisted in ventral view due to strong bilateral flattening. Colony composed of 8 zooids each with slightly displaced girdles. Sulci inconspicuous, on narrow edge of zooids, not continuous along the colony. Colony contains two nuclei and numerous chloroplasts although phagotrophy is said to occur. Nematocysts absent.

Size of colony: up to 60 μm in length.

Distribution: A littoral species found on wet sandy beaches around most of the British Isles. Original description from Port Erin.

Note: Other detailed descriptions of this species have been given by Balech (1956) and Dragesco (1965).

Polykrikos schwarzii Butschli

Fig. 14A, G

Butschli, 1873, p. 673, pl. 26, fig. 22.
Lebour, 1925, p. 67, pl. 10, fig. 2a.
Schiller, 1933, p. 550, fig. 580.

Colony composed of 2–8 units with half the respective number of nuclei. Shape an elongated ovoid with flattened anterior end and slightly depressed posterior end. Girdles of the zooids distinct and displaced by their own width, sulci appear to connect together to run the entire length of the colony. Lines of demarkation between zooids not very conspicuous. Cell contents pinkish (no chloroplasts) containing nematocysts and refractile granules. Nutrition heterotrophic.

Size: 100–150 μm long.

Distribution: A neritic species found all around the British Isles in the Summer. Reid (1978) has recently shown that the cysts thought to belong to this species can be found in the plankton around most of the British Isles.

Cyst: The cyst is spherical to extended ovoid, covered with a meshwork formed of protrusions from a relatively clear cell wall. Similar to a fossil cyst called *Valensia*:

Size: 80–160 μm long; 50–80 μm wide.

Family Lophodiniaceae
(incl. Crypthecodiniaceae)

AUREODINIUM Dodge

Dodge, 1967, p. 327

Motile cells broadly oval with deeply incised girdle, without lists; chloroplasts two or more, golden yellow bearing stalked pyrenoids; theca of delicate polygonal plates not visible by light microscopy except that it has a certain rigidity when the cell contents are extruded; flagella are inserted in the mid-ventral region, one longitudinal, other transverse; nucleus central. Coccoid phase present in old cultures with cells broadly oval, lacking theca and flagella.

Type species: *A. pigmentosum* Dodge

Aureodinium pigmentosum Dodge

Fig. 15L

Dodge, 1967, p. 329, fig. 1A.

Syn: *Gymnodinium pigmentosum* (Dodge) Loeblich, 1970, p. 903.

Motile cells broadly oval with hemispherical or flattened apices; girdle equatorial, deeply incised but without lists; sulcus slight. Two flagella inserted near centre of ventral surface, transverse flagellum encircling cell, longitudinal flagellum same length as cell. Cell clothed in delicate theca made up of membranes enclosing polygonal plates, fairly readily detached. Nucleus spherical, central, containing numerous small rod-shaped chromosomes. Chloroplasts two or more, much lobed, filling peripheral region of the cell and bearing stalked pyrenoids on their inner faces. Trichocysts and pusules absent.

Non-motile coccoid phase found in old cultures, lacks flagella and theca but otherwise has a similar cell structure, surrounded by a double membrane.

Size: c 10 μm long, c 7 μm wide.

Distribution: isolated from water samples taken from the western English Channel. May be common but due to small size has been collected only infrequently.

Note: The ultrastructure of this species was described in the original report (Dodge, 1967). It was notable that the main xanthophyll here is peridinin rather than the fucoxanthin found in a number of *Gymnodinium* species. Loeblich (1970), transferred this species to *Gymnodinium*, however, this move is resisted on the grounds of pigmentation and also on the presence here of delicate thecal plates. In fact, this clearly belongs to the family Lophodiniaceae rather than to Gymnodiniaceae (cf. Parke & Dodge, 1976).

CRYPTHECODINIUM Biecheler

Biecheler, 1952, p. 81.

Syn: *Glenodinium* Ehrenb., 1837 partim
Gymnodinium Stein, 1878 partim
Gyrodinium Kofoid & Swezy, 1921 partim

Vegetative cells spherical to ellipsoidal with an entire wall. Reproduction by autospores and zoospores. Zoospores ellipsoidal with a left-handed girdle which is incomplete on the ventral side. Sulcus situated on the left side of the epicone, slightly angled to the right. Cell covered with a delicate theca which can only be studied after careful staining. No chloroplasts at any stage but cells may have a slight pink colouration.

Type species: *Crypthecodinium cohnii* (Seligo) Chatton in Grasse.

Crypthecodinium cohnii (Seligo) Chatton in Grasse

Fig. 15I, J

Chatton, 1952, p. 347, fig. 251.

Syn: *Glenodinium cohnii* Seligo in Cohn, 1887, p. 156, figs. 22–26.
Gymnodinium fucorum Kuster, 1908, p. 352, figs. 1–4.
Gyrodinium fucorum Kofoid & Swezy, 1921, p. 305, fig. EE3.
Gyrodinium cohnii Schiller, 1933, p. 467, fig. 498.
Crypthecodinium setense Biecheler, 1938, p. 9, figs. 1–5.

Vegetative cells coccoid, non motile, spherical or ellipsoidal. Bounded by a firm periplast and containing numerous refractile granules but no chloroplasts. Reproduction by autospores and zoospores formed within the parent wall.

Size: 9–30 μm diameter.

Zoospores ellipsoidal with an incomplete left-handed girdle. Sulcus extending from the start of the girdle almost to antapex. Covered by a delicate theca the plate pattern of which has been revealed by silver staining (Biecheler) as 4', 3a, 5''; extra plate between end of girdle and sulcus; 5''', 3''''. Cell contains numerous refractile granules and an anteriorly placed nucleus.

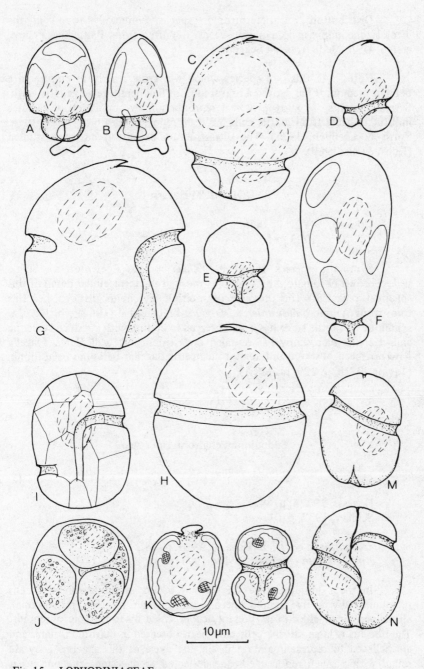

Fig. 15 LOPHODINIACEAE
A. *Katodinium rotundatum* B. *Katodinium rotundatum* (after Throndsen 1969) C. *K. glandalum* D. *K. asymmetricum* E. *K. fungiforme* F. *K. nieuportense* G. *Herdmania litoralis* ventral view H. *H. litoralis* dorsal view I. *Crypthecodinium cohnii* J. *Crypthecodinium cohnii* cyst K. *Endodinium chattonii* L. *Aureodinium pigmentosum* M. *Sclerodinium calyptroglyphe* dorsal view N. *Sclerodinium calyptroglyphe* ventral view.

Size: 10—27 µm long; 8—21 µm wide.

Distribution: A heterotrophic species commonly found in the littoral zone amongst decaying seaweeds. Reported from: Baltic Sea, France, eastern USA, Mediterranean Sea, etc.

Note: As this is a species which will grow readily in culture in a defined medium it has proved a useful tool for biochemical and genetic work. Several mutants are known. Sexual reproduction by copulation of gametes indistinguishable from motile cells has been reported (Beam & Himes, 1974; Tuttle & Loeblich, 1975). The ultrastructure of mitosis has been studied (Kubai & Ris, 1969).

ENDODINIUM Hovasse

Hovasse,   1922a, p. 845.
             1923, p 108.

Symbiotic species found in *Velella velella* (Coelenterata). Single yellow-brown chromatophore obscures most of the intracellular detail of this spherical organism. The presence of multiple pyrenoids distinguishes this form from *Symbiodinium microadriaticum* Freudenthal (1962) which is also symbiotic. A motile stage has been observed in culture with a reduced epicone and electron microscopy has revealed the existence of wall plates. Flagella have not been observed but some evidence of flagellar bases has been found. (Taylor, 1971b, p. 229, pl. 111A).

Type species: *E. chattonii* Hovasse.

Endodinium chattonii Hovasse

Fig. 15K

Hovasse, 1922b, p. 845.
Schiller, 1937, p. 61.

Syn: *Zooxanthella chattonii* Hovasse, 1924, p. 38.
        Amphidinium chattonii Taylor, 1971, p. 51, 227, pl. 1—3.

In the symbiotic stage this species occurs as single spherical cells surrounded by a five-layered theca (visible under electron microscope). Single peripheral chloroplast present accompanied by four ovoid pyrenoids. The nucleus is large. Motile cells have been observed in cultured isolates and are believed to represent a stage in the life cycle of this species. They are round with a much reduced epicone deflected to one side. The nucleus is posterior. The motile stage carries many features normally associated with *Amphidinium* and may indicate that the symbiotic habit is an unreliable character which may be obligate or facultative, (Taylor, 1971a). The presence of thecal plates would however suggest that this species is distinct.

Size: Symbiotic stage 12–14 µm diameter
             Motile stage 12–13 µm

Distribution: Symbiotic stage found in *Vellella velella* taken from English Channel, Mediterranean and off Florida.

Note: Originally described as the type species of the new genus *Endodinium* this organism was later transferred by Taylor (1971b) to *Amphidinium* on account of the shape of the symbiotic cells. This change was not accepted by Sournia, Cachon & Cachon (1975) because of the presence of plates in the theca.

HERDMANIA Dodge

Dodge. 1981.

Syn: *Gymnodinium* Stein 1878, partim.

Cell rounded in ventral view with a distinct apical notch; flattened dorso-ventrally; girdle median, incomplete on the ventral side of the cell; sulcus on left side of hypocone, broadening towards the antapex. Cell contains no chloroplasts but may contain refractile granules; covered with a delicate theca consisting of several plates.

Type species: *H. litoralis*

Named in honour of Catherine Herdman, the pioneer investigator of dinoflagellates living on sandy beaches.

Herdmania litoralis Dodge

Fig. 15G, H

Dodge 1981 (Br. phycol. J., 16)

Syn: *Gymnodinium agile* Herdman 1922, p. 29.
                     Herdman 1924, p. 81, fig 25.
                     Lebour 1925, p. 40, fig 11c.
        Gymnodium sp. Baillie 1971, p. 46, fig 5.

Non: *Gymnodinium agile* Kofoid & Swezy 1921, p. 184, fig Y9.

Cell strongly dorso-ventrally flattened, rounded in ventral view with a distinct apical notch. Epicone slightly larger than hypocone, hemi-circular with apical projection which lies to the left. Hypocone rounded with antapex sometimes indented by sulcus. Girdle just anterior to the median line but is not complete on the ventral surface, deep with overhanging borders, terminating in a posteriorly directed curve. Anterior flagella pore at junction of girdle and sulcus, longitudinal flagellum pore approximately 1.5 girdle widths behind anterior pore. Sulcus on left side of hypocone running from girdle to antapex, deep, widening posteriorly. The nucleus is situated in the epicone. One or two pusules may be present in the hypocone. The cell contents are clear and chromatophores are absent but refractile granules may be present. This species possesses a theca composed of delicate plates.

Size: length 25–35 µm; width 21–34 µm.

Distribution: Found on sandy beaches at Port Erin; Dorset, Sussex, Kent, S. Wales, Yorkshire, S.W. Scotland. Also from near Vancouver, Canada.

Note: There has been confusion over this species. Kofoid & Swezy (1921) described from the beach at La Jolla, California, an organism (*Gymnodinium agile*) with very much the same shape as the present species, but possessing orange-green chromatophores. They also showed the girdle to be complete and the sulcus median. Herdman (1922, 23, 24) soon found a similarly shaped organism on the beach at Port Erin. However, although she thought that the girdle was complete she did note the absence of chromatophores and the asymmetric sulcus. Baillie (1971) appears to have found both types on the sandy beaches of British Columbia. We have found the Herdman form from most parts of the British Isles. It clearly is a distinct species from that described by Kofoid & Swezy and furthermore, because it has a delicately armoured theca, it cannot remain in the genus *Gymnodinium*. The presence of the apical notch and the thin thecal plates suggest a possible affinity with *Katodinium* but the girdle is too far forward for that. There is one other genus which has an incomplete girdle and very delicate thecal plates, *Hemidinium* However, here the girdle traverses scarcely half the circumference of the cell and does not reappear on the ventral side. Hence it is necessary to create the new genus *Herdmania* with a new specific name *litoralis* to indicate its habitat.

KATODINIUM Fott

Fott, 1957, p. 287.

Syn: *Massartia* Conrad, 1926, p. 70.
Amphidinium Claparede & Lachman, 1858 partim
Gymnodinium Stein, 1878 partim

Non: *Massartia* Wildemann, 1897. (Zygomycetes)

Small, often fast-moving organisms which characteristically have a much longer and a wider epicone than hypocone. The girdle is usually horizontal and often formed by the difference in width between hypocone and epicone. The sulcus is normally only found on the hypocone. The cell is covered by a delicate theca which containes thin plates and ensures that normal fixatives preserve these organisms (require electron microscopy for confirmation of these). Many species are heterotrophic but some contain chloroplasts.

Type species: *Katodinium nieuportense* (Conrad) Loeblich & Loeblich.

Note: The taxonomic history of this genus is rather confused, in part owing to its origin from former members of the genera *Amphidinium* and *Gymnodinium*, and also due to the establishment of the invalid genus *Massartia* by Conrad (1926). Further confusion came when a new genus, *Katodinium*, was validly established by Fott (1957) but the species he proposed moving into it were not complete with basionym references. Thus it was left for Loeblich (1965) to complete the transfer of the species to *Katodinium*.

In the present work we remove *Katodinium glaucum* (Lebour) to *Gyrodinium*.

KEY TO THE SPECIES OF KATODINIUM

1a	Yellow-brown chloroplasts present	2
b	No chloroplasts present although refractile and orange bodies may be present	3
2a	Cell an inverted top-shape with a conical epicone	*K. rotundatum*
b	Cell ellipsoidal with a rounded apex	*K. nieuportense*
3a	Cell not appreciably flattened, apex rounded hypocone asymmetric. Small, less that 15 μm long	*K. fungiforme*
b	Cell somewhat dorso-ventrally flattened, apex rounded or notched. Size normally over 15 μm long	4
4a	Epicone two thirds of an ellipsoid, hypocone small and narrower than the epicone, deeply grooved by sulcus. Size less than 20 μm long	*K. asymmetricum*
b	Epicone half of an ellipsoid, hypocone almost as wide as epicone, rounded. Size generally more than 20 μm long	*K. glandulum*

Katodinium asymmetricum (Massart) Loeblich

Fig. 15D

Loeblich, 1965, p. 16.

Syn: *Gymnodinium asymmetricum* Massart, 1900, p. 82.
G. asymmetricum Massart, 1920, p. 132, fig. 22.
Massartia asymmetrica Schiller, 1933, p. 435, fig. 460.
M. asymmetrica Carter, 1937, p. 59, pl. 8, figs. 17–18.

Cell broadly ellipsoidal, slightly dorso-ventrally flattened, epicone rounded with apical notch, twice as large as the hypocone; epicone small and almost cleft into two by the asymmetrically placed sulcus (on the right side). Girdle more or less horizontal (may be offset about ½ its own width) with transverse flagellum encircling cell; longitudinal flagellum as long as cell. Nucleus situated in hypocone or behind girdle; no chloroplasts present; refractile or orange body may be present in epicone.

Size: length 10–22 µm; breadth 8–20 µm.

Distribution: A species of wet sands and salt marshes. Found in most of this type of sampling station around the S and W of the British Isles. Also reported from Belgium, France, eastern U.S.A.

Note: It is not easy to distinguish this organism from *K. fungiforme* except on the very marked asymmetry of the hypocone due to the position of the sulcus. The early descriptions reported that the epicone was entire but later workers have noted an apical notch. In shape and habitat this species also resembles *K. glandulum* from which it can only be distinguished by its smaller size range.

Katodinium fungiforme (Anissimowa) Loeblich

Fig. 15E

Loeblich, 1965, p. 16.

Syn: *Gymnodinium fungiforme* Anissimowa, 1926, p. 192, fig. 10.
G. fungiforme Schiller, 1933, p. 359, fig. 365.
Massartia fungiforme Schiller, 1955, p. 47.
Katodinium fungiforme Fott, 1957, p. 288.

A very small organism, scarcely flattened and with the appearance of a partly opened mushroom. The epicone is large and rounded, the hypocone reduced and may be slightly asymmetric. The girdle is horizontal and the sulcus slightly obliquely positioned on the hypocone. Transverse flagellum encircles cell, longitudinal flagellum about 1½ times cell length. The nucleus is situated in the hypocone and there are no chloroplasts or stigma. A refractile body (or bodies) may be present in the epicone.

Size: length 11–15 µm, breadth 8–12 µm.

Distribution: Frequently found in wet sand and salt-marsh pools, although easily overlooked because of its size. English Channel, Galloway, Lewis; also reported from saline pools in Russia, Czechoslovakia, France, etc.

Katodinium glandulum (Herdman) Loeblich

Fig. 15C

Loeblich, 1965, p. 16.

Syn: *Gymnodinium glandula* Herdman, 1924, p. 81, figs. 30–31.
G. glandula Lebour, 1925, p. 41, fig. 11g.
Massartia glandula Schiller, 1933, p. 435, fig. 461.

Body ovoid, slightly dorso-ventrally flattened. Epicone making up two thirds of the cell length, helmet-shaped with the apex distinctly notched. Hypocone short and not as wide as epicone. Girdle horizontal, deeply incised; sulcus running a short way onto epicone and on the hypocone obliquely to the right, up to the antapex. Transverse flagellum encircling cell, longitudinal flagellum 1–2 times cell length. Nucleus situated behind the girdle or in the centre of the cell; no chloroplasts; pale green, orange and other bodies may be present in cytoplasm.

Size: length 17–35 μm; breadth 14–28 μm.

Distribution: In wet sand, found around most of the British Isles.

Katodinium nieuportense (Conrad) Loeblich & Loeblich

Fig. 15F

Lieblich Jr & Loeblich III, 1966, p. 37.

Syn: *Massartia nieuportensis* Conrad, 1926, p. 70, pl. 1, fig. 1.
Schiller, 1933, p. 440, fig. 466.

Epicone ellipsoidal with squarish anterior end, up to ten times the length of the hypocone which is a small rounded button-like structure only about half the width of the epicone. The girdle is horizontal and no sulcus has been noted. The transverse flagellum encircles the cell and the longitudinal flagellum is about the same length as the cell. The nucleus is situated in the middle of the epicone and there are 2–4 oval yellow chloroplasts. Small oil droplets are visible in the cytoplasm.

Size: length 28–37 μm.

Distribution: Present in large numbers in brackish water near Nieuport (Belgium) in the summer of 1922, colouring the water yellow-brown (Conrad, 1926).

Katodinium rotundatum (Lohmann) Loeblich

Fig. 15A, B

Loeblich 1965, p. 16.

Syn: *Amphidinium rotundatum* Lohmann, 1908, p. 261, pl. 17, fig. 9.
Massartia rotundata Schiller, 1933, p. 438. fig. 464.
Gymnodinium minutum Lebour, 1925, p. 45, pl. 5, fig. 4.
Amphidinium pellucidum Redeke, 1935, p. 391, fig. 1.
A. redekei Conrad & Kufferath, 1954, p. 109.
Katodinium minutum Sournia, 1973, p. 44.
Massartia rotundatum var *Conradi* Kufferath, 1954, p. 108, pl. 7, fig. 9.

Body top-shaped, not flattened. Epicone twice to three times as long
as hypocone, conical with rounded or pointed apex. Hypocone narrower than
epicone, rounded. Girdle horizontal, a shallow depression; sulcus not clearly
observable. Transverse flagellum as long as the cell. Nucleus round, central
in position; yellow-brown chloroplasts present either two, one being band-
shaped around epicone the other in the bottom of the hypocone, or a variable
number peripherally placed. Orange body may be present in hypocone. Cell
covered by a delicate theca which contains very thin plates (Dodge & Crawford,
1970).

Size: length 8–17 μm, width 6–12 μm.

Distribution: A fairly common species of salt marshes and brackish
waters, but easily overlooked due to its small size. Found all around the
British Isles and also reported from Belgium, Norway, Baltic Sea, White Sea,
Eastern U.S.A., Adriatic Sea.

Notes: As a small species it has caused many taxonomic problems.
It would seem that the taxa *Gymnodinium minuta* of Lebour and *"Amphidinium (Rotundinium) pellucidum"* of Redeke rightly belong here as established and discussed by Conrad and Kufferath (1954). The var *Conradi* of
Kufferath appears to be merely a large form.

SCLERODINIUM Dodge

Syn: *Gyrodinium* Kofoid & Swezy, 1921 partim

Cell with spiral girdle, displaced about one third of body width, and
sulcus extending from apex to antapex. The cytoplasm is colourless and
contains no chloroplasts. A tough periplast or theca covers the cell.

Type species: *Sclerodinium calyptroglyphe* (Lebour) Dodge

Note: The new generic name describes the tough or hard cell covering found on the type species which contrasts with the weak theca typical of
Gyrodinium.

Sclerodinium calyptroglyphe (Lebour) Dodge

Fig. 15M, N

Dodge, 1981 (Br. Phyc. J. 16).

Syn: *Gyrodinium calyptroglyphe* Lebour, 1925, p. 52, pl. 7, figs. 3a and 6.
       G. calyptoglyphe Schiller, 1933, p. 448, fig. 477a, b.
       G. calyptoglyphe Drebes, 1974, p. 121, fig. 99a, b.

Cell small and irregularly ovate. Epicone with convex sides, slightly flattened or indented apex, notched by sulcus. Hypocone larger than epicone, sides straight to convex; antapex with a wider indentation formed by the sulcus than that in the epicone. The girdle is deep, wide, displaced about one third of the body width and makes about 1.25 turns in the left-handed spiral. The sulcus extends from the apex to the antapex and is very narrow on the epicone. Where it meets the girdle there is a small tongue-like process overlapping it from the right and the sulcus is obscured by a further larger tongue from the right in the intercingular region and yet another in the postcingular region. Thus the sulcus is hidden for most of its length but antapically it is much broader than apically. The nucleus is posterior and there are no chromatophores. The plasma is colourless but contains refractive granules and food vacuoles. A fairly tough periplast surrounds the cell and this may be poroid.

Size: 20–30 μm long; 15-23 μm wide.

Distribution: Occurs in summer all around the coasts of the British Isles and has been found in the central Atlantic.

Note: This organism preserves very well. It must therefore be suspected of possessing thecal plates or a tough cell covering and therefore cannot remain in the genus *Gyrodinium*. As regards shape it would appear to be not far removed from *Crypthecodinium*. In view of the uncertainty about the structure of the theca and about its life history a new genus has recently been created for this organism.

Sclerodinium striatum Dodge

Dodge, 1981, (Br. Phyc. J. 16) figs. 5, 8.

Syn: *Gyrodinium calyptroglyphe* Elbrachter, 1970, p. 12, fig. 31–34.

Cell small, ovoid with bluntly-pointed apex and slightly incised antapex. Girdle and sulcus as in *S. calyptroglyphe*. Epicone striated with ridges which may affect the anterior part of the hypocone. Chloroplasts absent.

Size: length 12–30 μm, width 10–23 μm.

Distribution: North Sea, Atlantic off N.W. Africa.

ZOOXANTHELLA Brandt

Brandt, 1881.

Syn: *Symbiodinium* Freudenthal, 1962.
        Gymnodinium Stein, 1878 partim.

Dinoflagellates with a main phase consisting of coccoid cells, with an entire wall, living symbiotically within the tissues of animals such as coelenterates and foraminiferans. There is also a free-living stage consisting of gymnodinioid cells covered with delicate thecal plates.

Type species: *Z. nutricula* Brandt.

Zooxanthella microadriatica (Freudenthal) Loeblich & Sherley

Loeblich & Sherley, 1979, p. 202.

Syn: *Symbiodinium microadriaticum* Freudenthal, 1962, p. 62, figs. 1–18.
        Gymnodinium microadriaticum Taylor, 1971, p. 232.

Coccoid cells found within coelenterates such as the sea anemone *Anemonia sulcata*. Entire wall; parietal chloroplast with several stalked pyrenoids; central nucleus. Free-living stage consists of gymnodinioid cell with delicate thecal plates having a complex and rather varied pattern (see Loeblich & Sherley, 1979) with the tabulation 5 ',5–6 a, 9–10 ", c20 c, 8–9 s, 7–8 '", 3 '"".

Size: Coccoid cell: about 13 μm diameter
         Motile cell: 12-14 μm long; 10–13 μm wide

Distribution: Probably world-wide.

Note: This very common symbiotic organism has had a rather chequered taxonomic history. See Loeblich & Sherley (1979) for the reasons for reverting to the genus *Zooxanthella*.

Order NOCTILUCALES

Family Noctilucaceae

KOFOIDINIUM Pavillard

Pavillard, 1928, p.l, fig.A,B.

Large dinoflagellates, non-pigmented in the mature state, strongly compressed laterally. Consisting of a hemi-circular keel or velum formed from the hypocone and constructed with a thick wall. Covering the epicone and girdle, which is on an apical crest, is a transparent hemi-sphaerical shell which is very loosely attached.

Studies by Cachon & Cachon (1967) have shown that this genus has a complex life-cycle in which spores first develop into *Gymnodinium*-like motile cells, then become *Amphidinium*-like, eventually enlarge to take on the typical *Kofoidinium* form and may end up looking like *Pomatodinium* as they produce spores. Some stages appear to be pigmented, possibly with lipid bodies or zooxanthellae rather than chloroplasts. Several of these stages have been described as independent species: *Gymnodinium pseudonoctiluca* Pouchet, *G. fulgens* Kof. & Swezy, *G. lebourii* Pav. and *K. velleloides* Pav.

Type species: *K. velleloides* Pavillard

Note: To date four species of *Kofoidinium* have been described, in addition to the numerous stages. It is very difficult to distinguish between the species other than by the size of the mature stages. On this alone it would appear that those we have found to the north of the British Isles belong to *K. velleloides*. We have also seen many examples corresponding to stages C and D of Cachon & Cachon (1967).

Kofoidinium velleloides Pav.

Fig. 16E, F

Pavillard, 1928, p.l, figs. A and B.
Cachon & Cachon, 1967, p. 440, fig. 6b.

Relatively large, transparent species with a rounded body and distinctive keel-like velum. The epicone is reduced and consists of a narrow girdle the left side of which is raised to become anterior to the rest of the epicone. The hypocone is rounded and contains the large nucleus slightly anterior of the centre. The distinctive keel-like velum is bordered by a narrow differentiated band and is rounded or irregular in preserved specimens. There is no tentacle, distinguishing it from *Noctiluca scintillans* and *Spatulodinium pseudonoctiluca*. Development stages have *Gymnodinium* and *Amphidinium*-like forms and may be pigmented.

Size: 100–300 μm wide.

Distribution: The present known distribution suggests a complementary occurrence to that of *Noctiluca scintillans* although there is a small zone of overlap. *Kofoidinium* has not as yet been found south of 54°N around the British Isles but since most members of the genus have been described from warm waters, notably the Mediterranean, it may be that this is a geographically distinct form similar to *K. arcticum* described from the Canadian Arctic Ocean (Bursa, 1964).

NOCTILUCA Suriray ex Lamarck

Suriray ex Lamarck, 1816, p. 470.

Non-thecate form which is globular and possesses a hydrostatic system so that its density may be regulated enabling it to occur over wide depths, usually the upper 50 metres, or may be buoyant and occur at the surface. Not differentiated in vegetative stage into epitheca and hypotheca; girdle not discernible but sulcus well developed as an oral groove. Cell large with striated tentacle, no flagella except in gametocyte generation where there is only one.

Type species: *N. scintillans* (Macartney) Ehrenberg.

Noctiluca scintillans (Macartney) Ehrenberg

Fig. 16 A–C

Kofoid & Swezy, 1921, p. 407, figs. KK 1–6.
Lebour, 1925, p. 69, figs. 17a–d.
Taylor, 1976, p. 186, pl. 39, figs. 478, 479.

Syn: *Medusa marina* Slabber, 1771, p. 67, pl. 8, figs. 4–5.
M. scintillans Macartney, 1810, p. 264, pl. 15, figs. 9–12.
Mammaria scintillans Ehrenberg, 1834, p. 559.
Noctiluca marina Ehrenberg, 1834, p. 559.
N. miliaris Suriray ex Lamarck, 1816, p. 470.
        1836, p. 1–16, pl. 1–2.
        Schiller, 1933, p. 553, figs. 582a–e.
        Zingmark, 1970, p. 122, figs. 1–18.
        Drebes, 1974, p. 127, figs. 108–109.

Cell large, inflated and reniform or subspherical, not differentiated into epitheca and hypotheca. The large nucleus is situated near the oral groove (sulcus) and cytoplasmic strands extend from it to the periphery of the cell. A prominent striated tentacle extends posteriorly. Numerous food vacuoles, often containing diatoms, are visible throughout the cytoplasm. Chloroplasts are absent and the cytoplasm is colourless, pink or the presence of green flagellated symbionts may give the cell a green colour. The latter colour has been reported from warm waters of the eastern Pacific (see e.g. Sweeney 1971). The species is luminescent. Reproduction is by fission or gametogenesis. Zingmark (1970) describes the production of uniflagellate gametes and also gives a discussion on the life cycle. The species can occur in large numbers off the southern coast of Britain and it has been reported as the causative organism of red tides in many parts of the world (see e.g. Fung & Trott, 1973 from Hong Kong).

Size: 200—2,000 µm diameter.

Distribution: Not seen north of 56°N around Britain but commonly found south of this line near shore and estuarine. Elsewhere widespread in distribution in temperate and tropical waters.

Note: This species is frequently referred to as *N. miliaris* although Macartney's specific name has priority. Taylor (1976) suggests that the simplest solution to the problem of nomenclature is to accept the priority of *scintillans* especially as this has been used by two major works: Kofoid & Swezy (1921) and Lebour (1925); Schiller (1933) together with many recent authors have preferred *N. miliaris*.

SPATULODINIUM Cachon & Cachon

Cachon & Cachon, 1967, p. 441, fig. 9.

Having a laterally flattened mature stage with girdle running along the anterior margin with extended sulcus reaching the dorsal side of the cell. Cell rounded in lateral view with unstriated tentacle extending from the ventral surface. No velum as in *Kofoidinium*. A curved lip is characteristically present anterior to the tentacle. A juvenile gymnodinioid stage occurs with differentiated epicone, girdle and hypocone.

Type species: *S. pseudonoctiluca* (Pouchet) Cachon & Cachon.

Spatulodinium pseudonoctiluca (Pouchet) Cachon & Cachon

Fig. 16D

Cachon & Cachon, 1967, p. 441, fig. 9.

Syn: *Gymnodinium pseudonoctiluca* Pouchet, 1885, p. 71, pl. 10,
                                      figs. 34—37.
                                      Paulsen, 1980, p. 97, fig. 135.
                                      Kofoid & Swezy, 1921, p. 243.
                                      Lebour, 1925, p. 44, figs. 12a, b.
                                      Schiller, 1933, p. 399, figs. 417b—c.

This species is characterized by a very small epicone and in the notilucoid or mature stage a long unstriated tentacle projecting at a right angle to the main axis of the body, in the sulcal region. The cell is oval or round in shape, laterally compressed with a differentiated border. The reduced epicone may be a little narrower than the hypocone and is rounded or flattened anteriorly. The cingulum is almost enveloped so that the transverse flagellum

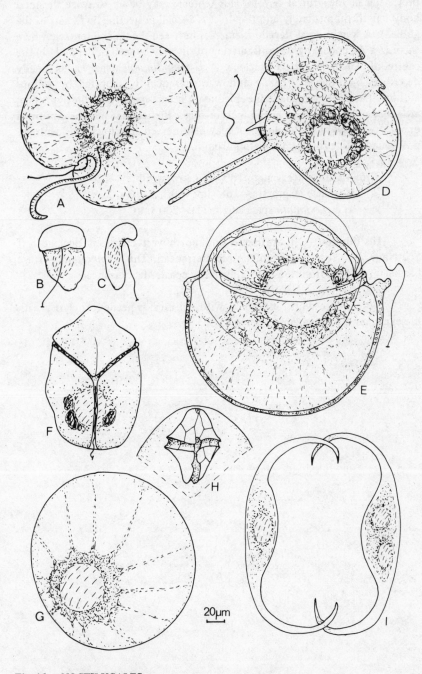

Fig. 16 NOCTILUCALES

A. *Noctiluca scintillans* B.C. *Noctiluca scintillans* gametes (not to scale)
D. *Spatulodinium pseudonoctiluca* E. *Kofoidinium velleloides* sporont
F. *Kofoidinium velleloides* developmental stage

PYROCYSTALES

G. *Pyrocystis noctiluca* H. *Pyrocystis noctiluca* thecate swarmer formed
within cyst I. *P. hamulus* pair of cells

lies in a tube-like structure. The hypocone is rounded and has the tentacle projecting in the ventral region. This tentacle may be up to twice the total body length. Immediately anterior to it is a small projecting lip (a part of the girdle?). A longitudinal flagellum emerges between the lip and tentacle where there is a deep sulcal groove. Radiating fibrils may be seen in the body in the centre of which is a large nucleus. A *Gymnodinium* stage was originally described by Pouchet (1885) and was subsequently reported by various authors. The girdle, sulcus and the epicone, although smaller than the hypocone, are more prominent than in the mature stage; the tentacle may be unformed or various stages in its development may be observed. The nucleus is centrally positioned with cytoplasm radiating from it.

Size: Mature stage: length 100−163 µm; width 89−120 µm.
Tentacle length 100−168 µm; width 5−10 µm.
Gymnodinium stage length 110−200 µ m.

Distribution: Off north-east and north-west coasts of England, off Scotland. Also recorded from the Mediterranean. The *Gymnodinium* stage has been recorded from the Faroes, Brittany coast, Mediterranean.

Note: More work, especially with cultures, is needed to clarify with life cycle and development of this species.

Order PYROCYSTALES

Family Pyrocystaceae

PYROCYSTIS Murray ex Haeckel

Haeckel, 1890, p.30.

Syn: *Gymnodinium* Stein, 1978, partim
      Murracystis Haeckel, 1891, p. 261.
      Diplodinium Klebs, 1912, partim
      Dissodinium Klebs in Pascher, 1916, partim

Marine dinophytes with a dominant coccoid stage which is fusiform, lunate or spherical. Cell wall smooth and unsculptured, cytoplasm parietal or clustered around the nucleus, chloroplasts numerous. Normally biolumine-scent. Asexual reproduction by the formation of 1 or 2 swarmers which may be thecate and *Gonyaulax*-like or athecate and *Gymnodinium*-like. After release the swarmers swell up and form a new coccoid stage. Mainly found in warm waters.

Type species: *P. noctiluca* Murray ex Haeckel.

Note: There has been considerable confusion and taxonomic chaos between this genus and *Dissodinium*. This has now been thoroughly sorted out by Elbrachter and Drebes (1978). Other contributions to our knowledge of the genus will be found in Taylor (1972, 1976), Swift and Durbin (1971), Swift & Wall (1972) and Meunier & Swift (1977).

Pyrocystis noctiluca Murray ex Haeckel

Fig. 16G, H

Haeckel, 1890, p.30.

Syn: *P. pseudonoctiluca* Wyville-Thomson in Murray
                  Schiller, 1937, p.485, fig. 556.
      Goniodoma concava Gaarder, 1954, p.27, fig. 32.
      Gonyaulax concava Balech, 1967a, p.108, pl.6. figs. 108—116.

A globular coccoid dinophyte which is sparsely filled with cytoplasm, most of which is clustered around the nucleus. A motile thecate stage is produced which has delicate plates in the tabulation 4', Oa, 6", 6c, 6"', 1p, 1"" which is correct for the genus *Gonyaulax*. Smaller globose cysts may form within the large ones. Strongly bioluminescent.

Size: coccoid cyst, 350—800 μm diameter
      thecate stage, 60—80 μm diameter

Distribution: A warm-water oceanic species which may occasionally be found to the west of the British Isles.

Note: Details of its structure and life history are to be found in Swift & Durbin (1971) and Taylor (1972).

Pyrocystis hamulus Cleve

Fig. 16I

Cleve, 1900, p.19, pl.7, fig. 23.
Schiller, 1937, p.490, figs. 563—565.
Taylor, 1976, p.180, figs. 446—450.

This is a distinctive organism in which the cysts are elongated and curved, with a central swelling containing the cytoplasm. It usually occurs in pairs which are held together by the attachment of the distal ends of the arms. The contents may be seen to be divided into two.

Size: c. 250 μm long.

Distribution: Said to be a tropical species but organisms attributed to this species were found in the English Channel and the Clyde.

Note: Little is known about this species except that it is very variable, further work is needed to show that it is related to the other species placed in this genus. Drebes 1981 *(Br. phycol. J.,* 16, 207*)* has described the germination of *P. hamulus*-like resting spores to give dinospores of *Dissodinium pseudolunula* (see p. 257).

Order PERIDINIALES

Family Pyrophacaceae

HELGOLANDINIUM von Stosch

von Stosch 1969b, p.576, fig 3.

Globular dinoflagellates with dense contents and thin thecal plates. Girdle median. Cells contain numerous chloroplasts and a U-shaped nucleus. Cell division takes place after shedding of the theca.

Type species: *H. subglobosum* von Stosch.

Helgolandinium subglobosum von Stosch

Fig. 17E, F

von Stosch, 1969b, p.576, figs. 1–4.
Drebes, 1974, p.130, fig. 111.

Cell more or less spherical but with broadly pointed epicone bearing a distinctive curved apical pore. Girdle median, deeply incised. Cell covered with delicate theca which readily becomes detached. Plate formula Po, 4', 9'', 10c, ?s, 7''', 3''''. Thecal plates smooth but perforated by numerous pores. Cell contains many chloroplasts and an elongated nucleus. Smooth-walled cysts form within the theca.

Size: 22–56 μm diameter

Distribution: Common in the North Sea, also found around most of the British Isles.

Note: First described by von Stosch (1969b) from Helgoland, this organism is now appearing in many samples from the British Isles. In the past it was presumably confused with *Gonyaulax*, *Heteraulacus* or *Glenodinium*. It has been suggested (Loeblich, in Drebes, 1974) that it should more correctly be ascribed to the genus *Fragillidium*.

PYROPHACUS Stein

Stein, 1883, pl.24.

Cell anterio-posteriorly flattened having a bi-convex lens shape when seen from the side. Rather thin thecal plates with a variable plate formula 5–9', 0–8a, 7–15'', 9–16c, 8–17''', 0–9p, 3–7''''. Cell contents tend to round up, pushing apart the epitheca and hypotheca.

Type species: *P. horologium* Stein.

Note: Until recently, this genus contained only one species with several varieties. It has been studied in some detail from Florida by Steidinger & Davis (1967) who described a new form termed 'B 1' with a very large number of plates. Subsequently, in a detailed study of thecae and cysts from many sources, Wall & Dale (1971) raised two of the varieties to distinct species and showed that the variety 'B 1' had a tuberculate cyst similar to one previously described from fossil deposits and known as *Tuberculodinium vancampoae.* To date only *P. horologium* has been found around the British Isles, although it is possible that the large tropical species *P. steinii,* which has a larger number of antapical plates, might be brought here by the North Atlantic Drift.

Pyrophacus horologium Stein

Fig. 17A, B

Stein, 1883, p.28, pl.28, figs. 6–13.
Lebour, 1925, p.139, pl.29, figs. 4a–c.
Schiller, 1937, p.87, figs. 73–74.
Drebes, 1974, p.130, fig. 110.

Cell discoidal, almost circular in anterior view but biconvex lens-shaped in lateral, dorsal or ventral views. Theca weakly sculptured with a large and variable number of plates giving a plate formula within the range: 5–6'. 0–1a, 7–10", 9c, 8–10'", 0–1p, 3–5"", Intercalary bands narrow to very broad. Cytoplasm with a strong tendency to round up, containing numerous chloroplasts. Reproduction by formation of 2–4 spores (? cysts); cysts are oblate reniform with a smooth transparent wall surrounded by gelatinous material, and contain brownish protoplasm with a reddish area.

Size: width 35–136 µm, height 32–125 µm

Distribution: Found all around the British Isles but most commonly near to land and in the North Sea.

Note: This species is rather variable as regards its plate number and arrangement but the form of the organism is quite distinctive.

Family Peridiniaceae

CACHONINA Loeblich

Loeblich III, 1968, p.92, figs. 4–7.

Syn: *Glenodinium* Ehrenberg, 1837, partim

Small ellipsoidal dinoflagellate with a very delicate thecal structure consisting of about 34 distinctive plates. Cell contains parietally arranged brown chloroplasts and a large spherical nucleus is situated in the hypocone. Thecal plates very delicate and variable even within cultures.

Type species: *Cachonina hallii* (Freudenthal & Lee) comb. nov.

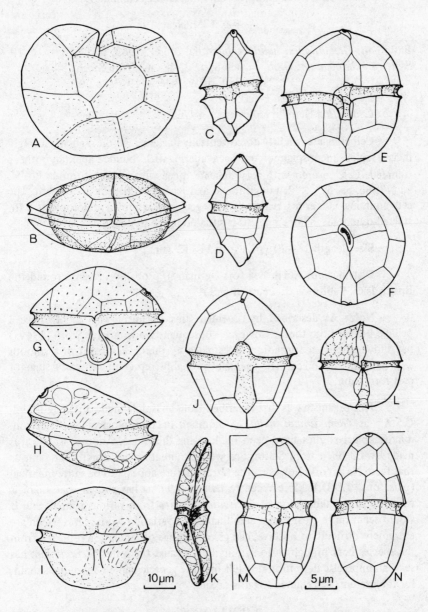

Fig. 17 A. *Pyrophacus horologium* (hypotheca) (after Wall & Dale 1971)
B. *Pyrophacus horologium* ventral view C. *Heterocapsa triqueta* ventral view
D. *Heterocapsa triquetra* dorsal view E. *Helgolandinium subglobosum* ventral
view (after Von Stosch 1969) F. *Helgolandinium subglobosum* epitheca
G. *Coolia monotis* ventral view H. *Coolia monotis* side view I. *Glenodinium
monensis* J. *G. foliaceum* ventral view K. *G. foliaceum* side view
L. *G. danicum* (after Elbrachter pers. com.) M. *Cachonina hallii* ventral view
(after Loeblich & Von Stosch) N. *Cachonina hallii* dorsal view (after Loeblich
& Von Stosch)

Cachonina hallii (Freudenthal & Lee) comb. nov.

Fig. 17M, N

Basionym: *Glenodinium hallii* Freudenthal & Lee, 1963, p.187, figs. 18, 19.
Syn: *Cachonina niei* Loeblich, 1968, p.92, figs. 4—7 von Stosh,
               1969a, p. 560, fig. 1. Balech, 1977a, p.60
               figs. 1—20.
     C. illdefina Herman & Sweeney, 1976, p 8.

Cell elongate, slightly dorso-ventrally flattened. Girdle median, strongly incised; sulcus inconspicuous; epitheca tapering with rounded apex; hypotheca rounded. Cell covered with very delicate theca with average formula Po, 5', 3a, 7", 6c, 5s, 5''', 2''''. Longitudinal and transverse flagella are present. The cell contains numerous parietally-arranged yellow-brown chloroplasts with stalked pyrenoids; a large nucleus is situated in the hypocone.

Size: length 17—20 μm; width 11—12 μm.

Distribution: North Sea (off Germany), probably present around the British Isles; California, E. coast of U.S.A.

Note: As described by Loeblich this organism was said to have a pellicle beneath the theca. However, subsequent work by von Stosh (1969a) has shown that cyst formation takes place by the development of a smooth wall beneath the theca. The 'pellicle' probably represents an early stage of cyst formation.

This organism was first described from an isolate made off Long Island, U.S.A., as *Glenodinium hallii*. A number of physiological and electron-microscopical studies have been made using this clone. Another isolate was made by Loeblich from Salton Sea with a tabulation not exactly the same as that from New York; however, the ultrastructure appears to be quite identical (pers. obs. J.D.D.). More recently Balech (1977a) has shown that there is variation in the tabulation of *Cachonina isolates* from various sources and it would seem quite clear that a combination of variation, with plates which are extremely difficult to observe, can give the range of tabulation reported from the species here brought into synonymy. Because the genus *Glenodinium* has such an imprecise definition it is felt that the generic name *Cachonina* should, for the present, be retained for this taxon.

COOLIA Meunier

Meunier, 1919, p.67.

Cell lens-shaped in ventral view with a pronounced oblique axis giving the appearance of being squashed and distorted from front to back. Girdle equatorial, excavated and just slightly left-handed, sulcus in hypotheca but not reaching antapex. Theca consisting of distinct plates thought to be some-what irregular. Cell contents dense and chloroplasts present.

Type species: *C. monotis* Meunier.

Coolia monotis Meunier

Fig. 17G, H

Meunier, 1919, p.68, pl.19, figs. 13–19.
Lebour, 1925, p.138, fig. 43.
Balech, 1956, p.47, figs. 53–66.

Syn: *Ostreopsis monotis* Lindemann, 1928, p.97.
                  Schiller, 1937, p.472, figs. 542a–d.

Glenodinium monotis Biecheler, 1952, p.44, figs. 20–22.

Cell compressed anterio-posteriorly with an oblique axis. Ellipsoidal in apical view. The epitheca is asymmetrical with the plates on the right larger than those on the left. There are 11 plates interpreted as 3', 7'' and a pore plate. For a detailed description of the plates see Balech (1956). The girdle is composed of six plates, is excavated deeply and slightly left-handed. Hypotheca, like epitheca, rounded in ventral view, with five precingular and two antapical plates. The latter were considered by Meunier & Lebour to be sulcal plates. The sulcus itself is deep and bordered by 1''', 5''', 2'''' and 1'''' and does not reach the antapex. There are numerous yellow-brown chromatophores which tend to radiate from the centre of the cell. The nucleus is large, dorsally situated in the hypocone. Two pusules are present which seem to open independently in the sulcus. Plate formula 3', 7'', 6c, 5''', 2''''. The plates have numerous pores scattered over them.

Size: 32–40 μm long.

Distribution: An organism of littoral situations. Found rarely around coast of Britain in semi-permanent pools, especially on salt marshes. Also from oyster beds at Nieuport, Belgium (Meunier, 1919), the Mediterranean and Brittany.

Note: It is possible that *Glenodinium obliquum* Pouchet (1883) is the same as this species or it may be closely related. This needs further study.

GLENODINIUM Ehrenberg

Ehrenberg 1837, p.174

Glenodinium was established for fresh water dinoflagellates which had a red stigma. Stein, 1883, used the name for those species with a firm homogeneous wall (i.e. with no plates). In 1925 Lebour defined the genus as "composed of epitheca, girdle and hypotheca, not divided into plates and not divided by a seam. Frequently, if not always, the theca, on treatment with reagents, shows a pattern of numerous small platelets, which apparently cannot be disassociated." Since then, species which could not confidently be included in (*Proto*)-*peridinium* because plates had not been seen, or *Gymnodinium* because the cell wall was clearly thickened have been placed tentatively in *Glenodinium*. More detailed examination of morphology has shown that most of the species in the genus bear very tenuous relationships to one another and taxonomic reappraisal is required to define the limits of the genus. In the meantime, workers have found it convenient to continue to use the genus for small species, with tabulation which can only be seen after the cell is empty, or by staining, with epitheca and hypotheca more or less equal and with a median or only slightly offset girdle. *G. danicum* has been shown to be covered with numerous polygonal plates whereas in *G. foliaceum* the plate formula is 3–4', 2a, 7", 5''', 2'''', and *G. monensis* has yet to have plates described. *Coolia monotis* which Biecheler (1952) included in *Glenodinium*, has the formula 3', 7", 6c, 5''', 2''''. Biecheler gave the tabulation for the genus as 3–5', 0–1a, 6–7", 5''' and 2'''' and hence placed her *G. foliaceum* in *Peridinium*. The situation may be helped when some of the firm-walled *Gymnodinium* and *Gyrodinium* species have been examined for plates, if any.

Types species: *G. cinctum* Ehrenberg (F.W.)

Glenodinium danicum Pauls

Fig. 17L

Paulsen, 1907, p.6, figs. 2a–h.
Lebour, 1925, p.86, pl.12, fig. 6.
Schiller, 1937, p.110, fig. 106.
Braarud, 1935, p.104, fig. 23.

Small, rounded cell with girdle and sulcus clearly visible. Epitheca with small apical horn, hypotheca sometimes with antapical spine at left border of sulcus. Girdle about one tenth total body length left-handed, displaced about one girdle width. Sulcus clearly marked on left side, narrow but broadening posteriorly and in some specimens continuing as far as apex. Nucleus central or in epicone. Plasma yellow, chromatophores not described but may be present. Since the original description plates have been seen in the cell wall (Braarud, 1935; Elbrachter, pers. comm.), these are small, polygonal, and form a reticulated pattern. Meunier (1910) found a specimen containing a cyst with two antapical spines, the contents emerged when the girdle came apart.

Size: 28–36 μm long.

Distribution: Found all round coast of Britain except English Channel. Elsewhere recorded from Skagerak, Baltic, Barents Sea, S. Africa.

Glenodinium foliaceum Stein

Fig. 17J, K

Stein, 1883, pl.3, figs. 22—26.
Paulsen, 1908, p.22, fig. 25.
Schiller, 1937, p.120, figs. 117a—m.
Prager, 1963, p.1—10, figs. 1—3.

Syn: *Kryptoperidinium foliaceum* Lindemann, 1924, p.114.
                            Lebour, 1925, p.104, pl.16, figs. 2a—f.
      Peridinium foliaceum Biecheler, 1952, p.78, figs. 47—49.

Cell markedly flattened dorso-ventrally with convex dorsal and
concave ventral surface. Epicone slightly larger than hypocone, rounded.
Hypocone flattened posteriorly. Girdle excavated without lists, slightly displaced. The sulcus is narrow and extends about half way down the hypotheca.
Longitudinal flagellum about the length of body. A large brick-red stigma is
situated centrally on the hypocone, usually near the flagella pore (Dodge &
Crawford, 1969). The nucleus is central or in the hypocone. Numerous discoid
greenish-brown chromatophores are randomly positioned in the cell or
arranged in a circle. Pusules are present. Thin walled cysts have been described
by Prager (1963). There has been some disagreement amongst authors over
the plate formula which is 3—4', 2a, 7", 5"', 2"'' and is therefore similar to
some *Protoperidinium* species. However, plate tabulation alone is not considered a definitive taxonomic criterion and the presence of a stigma, the
delicate nature of the plates and its habitat (brackish water) suggest that this
species should be separate from the genus *Protoperidinium* although the
genus *Glenodinium* is not satisfactorily defined.

Size: 14—50 µm long; 11—45 µm wide.

Distribution: Brackish water around Britain, frequently in large
numbers. Also recorded from the Baltic, Mediterranean and N.E. U.S.A.

Note: The presence of a second nucleus in this organism was established
both by light and electron microscopy (Dodge, 1971). More recently it has
been found that a very similar organism, currently called *Peridinium balticum*
also has two nuclei (Tomas *et al.*, 1973). The significance of this is that both
of these dinoflagellates appear to be harbouring a symbiotic chrysophyte to
which also belong the chloroplasts (Tomas & Cox, 1973). It is not surprising
therefore that in both species the main carotenoid pigment is fucoxanthin
(Jeffrey *et al.*, 1975).

Glenodinium monensis Herdman

Fig. 17I

Herdman, 1923, p.3, fig. 1.
Lebour, 1925, p.86, fig. 24d.
Schiller, 1937, p.121, fig. 118.

Cell rounded and flattened dorso-ventrally. No projections at the apices and epicone and hypocone positioned excentrically with former projecting slightly to the right and latter to the left. Girdle centrally positioned sometimes with a small displacement. The sulcus is very short and projects an equal distance onto the epicone. The longitudinal flagellum is about three times the length of the body. The ovoid nucleus is centrally positioned. Yellowish green colour to cell contents. No plates described.

Size: 23–25 μm length

Distribution: Recorded in sand from Port Erin, Isle of Man, and Bembridge, Isle of Wight.

HETEROCAPSA Stein

Stein, 1883, p.9, 13.

Small irregularly shaped dinoflagellate with rounded apex and narrowed antapex. Girdle median. Nucleus spherical, numerous brown chloroplasts and large pyrenoid present. Thecal plates relatively thin, difficult to determine, and show some variation:- Po, 4–6', 2–3a, 6–8", 6c, ? s, 5–6''', 0–1 p, 2''''.

Type species: *H. triquetra* (Ehrenberg) Stein

Heterocapsa triquetra (Ehrenb.) Stein

Fig. 17C, D, Pl. III a.

Stein, 1883, p.13, pl.3, figs. 30–40.

Syn: *Glenodinium triquetrum* Ehrenberg, 1840, p.200.
Properidinium heterocapsa Meunier, 1919, p.58, pl.19, gis. 43–49.
Peridinium triquetra Lebour, 1925, p.109, pl.18, fig. 2.
P. triquetrum Schiller, 1937, p.145, fig. 147.

Cell irregularly spindle-shaped. Epicone with straight sides and rounded or squared apex; hypocone variably shaped, often with displaced antapex. Girdle slightly excavated, median, fractionally depressed on the right side; sulcus inconspicuous, not reaching the antapex. No spines or ornamentation; thecal plates smooth with trichocyst pores; typical plate formula 5', 3a, 7", 6c, 7s, 5''', 1p, 2''''. Cell contents dense with a brown peripheral chloroplast, a large pyrenoid situated in a median position and nucleus in the epicone. Spiny cysts have been reported (Braarud & Pappas, 1951). A recent paper (Pennick & Clarke, 1977) has reported the presence of tiny scales over the surface of the cell.

Size: length 16–30 μm; width 9–18 μm.

Distribution: A neritic species which may produce large numbers in summer in coastal areas. Found all around the British Isles except the N. of Scotland. Also common along the coasts of N. Europe and found off the U.S.A.

THE DIPLOPSALIS GROUP

Small globular or lenticular dinoflagellates with the theca divided into plates. The theca is smooth or finely granular and an apical pore is present. The girdle is equatorial and there are prominent lists. A feature common to all members of the group is a pronounced list attached to the left side of the sulcus and often protruding as a hook at the posterior end of the cell. No chloroplasts are present and the cell plasma is a pinkish colour. Some species are known to form cysts.

The taxonomy of the species which, following earlier authors, are here treated as a 'group' is extremely confused. They definitely all belong to the order Peridiniales and to the family Peridiniaceae. They can be distinguished from the genus *Peridinium*, in its current usage, by having only three girdle plates as opposed to the five or six in that genus. Some can be distinguished from both *Peridinium* and *Protoperidinium* by having only one antapical plate and three apical plates, only one or two anterior intercalaries and only six pre-cingular plates.

The following is an attempt to give a brief historical background to the confusion of generic and specific names which have been applied to the organisms in this group. A detailed revision based on scanning electron microscopy (Dodge & Hermes, In preparation) has provided useful information to help determine the number of distinct genera in material from around the British Isles.

The first organism of this group to be described was *Diplopsalis lenticula* (Bergh, 1882). Unfortunately no tabulation was given and the first to appear in the press was Stein's (1883) plate labelled as *D. lenticula* but actually including at least two distinct organisms. In succeeding years there were several further descriptions of organisms supposed to be *D. lenticula* or forms of it and several other species were described. In 1910 Meunier formed the genus *Diplopsalopsis* for *Peridinium orbiculare* (cf. Paulsen, 1908) since this species had only three apical plates and two anterior intercalaries and seven pre-cingular plates. Mangin (1911) described the next member of the group, *Peridiniopsis asymmetrica,* using Lemmerman's (1904) genus. The type species of this genus has recently been shown by Bourrelly (1968) to have six girdle plates. In 1913 Pavillard devised the genus *Diplopeltopsis* for *Diplopsalis lenticula* f. *minor* but recently Loeblich & Loeblich (1970) have shown that the name had earlier been used for a lichen genus so the new name *Zygabikodinium* was proposed.

The confusion already existing was sorted out as far as possible by Lebour (1922) who clarified the description of four genera and also described a new tiny spherical dinoflagellate as *Peridiniopsis rotunda*. Other authors were not happy with the use of several genera and Lindemann (1928) placed all the species of the group in *Diplopsalis* but shortly after Schiller (1937) moved them again into the ill-defined genus *Glenodinium*. Abe (1941) formed the new genus Dissodium and placed in it the organism which had been described as *Peridiniopsis asymmetrica*. He also created another new genus *Lebouria*. Nie (1943) placed several known species in *Diplopsalis* and described some new ones including *D. excentrica*.

Balech (1964) introduced the new genus *Oblea* for *Peridiniopsis rotunda* and a new species. Loeblich (1970) suggested that the name *Diplopsalis* was invalid according to the Zoological Code as it was pre-occupied by a sub-genus of birds. He proposed substituting Abe's *Dissodium* for a number of species of *Diplopsalis* and *Peridiniopsis*. Taylor (1976) proposed the retention of *Peridiniopsis* and the abandonment of the confusing name *Dissodium*. Most recently Balech (1976d) has utilised as a generic name *Diplopelta*, an invalid name as there is no description of it although it has been used informally by Stein (1883) and as a sub-genus of *Peridiniopsis* by Lebour (1922).

In the account that follows the species found around the British Isles are placed in five genera which are arranged in alphabetical order. The name *Diplopsalis* is conserved as it is still valid under the Botanical Code and is essential for the preservation of a small element of continuity in a group of organisms which have had a remarkably turbulent taxonomic history. To identify these organisms it is essential to find a clear or empty cell so that the details of the tabulation can be worked out.

KEY TO THE GENERA AND SPECIES OF THE *DIPLOPSALIS* GROUP

1a	Cell with single antapical plate	2
b	Cell with two antapical plates	4
2a	Epitheca with only one, large, intercalary plate	*(Diplopsalis) D. lenticula*
b	Epitheca with one large and one very small intercalary plate	*(Zygabikodinium)* 3
3a	Cell lenticular	*Z. lenticulatum*
b	Cell globular	*Z. pseudooblea*
4a	First apical plate five-sided (meta)	*(Oblea) O. rotunda*
b	First apical plate four-sided (ortho)	5
5a	Seven precingular plates	*(Diplopsalopsis) D. orbicularis*
b	Six precingular plates	*(Dissodium)* 6
6a	Cell lenticular, plate 2a of similar size to apicals	*D. asymmetricum*
b	Cell globular, 2a very large and apical series of plates on ventral side of epitheca	*D. excentricum*

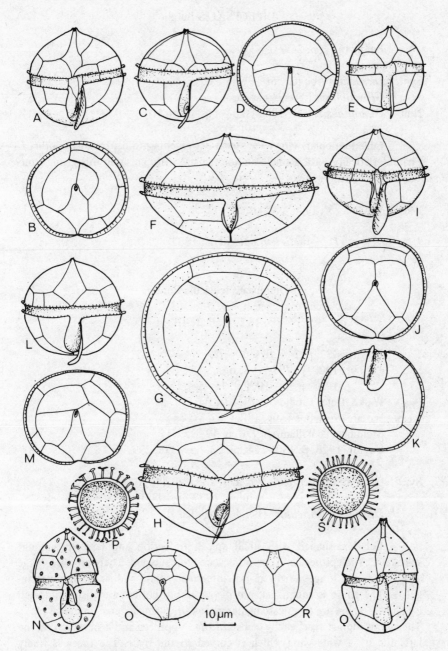

Fig. 18 "DIPLOPSALIS group"

A. *Oblea rotunda* B. *Oblea rotunda* apical view C. *Zygabikodinium lenticulatum* D. *Zygabikodinium lenticulatum* apical view E. *Z. pseudo-oblea* F. *Dissodium asymmetricum* G. *D. asymmetricum* apical view H. *D. excentricum* I. *Diplopsalis lenticula* J. *Diplopsalis lenticula* apical view K. *Diplopsalis lenticula* antapical view L. *Diplopsalopsis orbicularis* M. *D. orbicularis* apical view N. *Scrippsiella faeroense* O. *Scrippsiella faeroense* apical view P. *Scrippsiella faeroense* cyst (smaller scale) Q. *S. trochoidea* R. *S. trochoidea* antapical view S. *S. trochoidea* cyst (smaller scale)

DIPLOPSALIS Bergh

Bergh 1882, p. 244

Syn: *Dissodium* Abe partim
        Glenodinium Ehrenb. partim
Non: *Diplopsalis* Sclater 1866 (sub-genus of birds)

Cells lenticular with wide sulcal list reaching to centre of hypotheca. Characteristically with a single antapical plate distinguishing this genus from *Dissodium* which has two.

Plate formula: 3', 1a, 6", 3c, 6s, 5''', 1''''

Type species: *Diplopsalis lenticula* Bergh 1882

Diplopsalis lenticula Bergh

Fig. 18 I–K, Plate IIIb, c

Bergh, 1882, p. 244, figs. 60–62.
Lebour, 1922, p. 795–798, figs. 1–5
         1925, p. 99–100, pl. 15, figs. 1a–e.
Wall & Dale, 1968a, p. 279, text fig. 7.
Wood, 1968, p. 54, fig. 132.
Steidinger & Williams, 1970, p. 49, fig. 51.
Taylor, 1976, p. 130–131, pl. 28, figs. 298, 299.

Syn: *Glenodinium lenticula* Pouchet, 1883, p. 44, pl. 20–21.
                   Schiller, 1937, p. 103, figs. 95a–h.
     Dissodium lenticulum (Bergh) Loeblich III, 1970, p. 905.

Cell lens-shaped with small apical projection and large, prominent sulcal list. The epitheca and hypotheca are almost equal. There are three large apical plates, a long, narrow dorsal intercalary plate and six narrow precingulars. The girdle is equatorial, not displaced but bordered by lists supported by spines. There are five postcingular plates and a single, large antapical plate. The sulcus almost reaches the centre of the hypotheca and is bordered on the left side by a wide wing which is curved to the right. The theca is finely punctate and the plasma pink. Plate formula: 3', 1a, 6", 3c, 6s, 5''', 1''''.

Size: 23–48 µm long; 32–68 µm diameter.

Distribution: A common species found all round the coast of Britain predominantly in spring and summer. Elsewhere recorded from the Baltic, Mediterranean, Red Sea, Australia, Gulf of Aden, Indian Ocean, Caribbean, Antarctic and China Sea.

Cysts: The resting spores are thin-walled, smooth, brown and sphaerical. The archeopyle is hexagonal and situated dorsally at the position of 1a in the thecate stage. The cysts are not known in the fossil record (Wall & Dale, 1968a).

Cyst size: 49–57 μm diameter.

Note: Frequently recorded collectively with *Dissodium asymmetricum* and *Zygabikodinium lenticulum*.

DIPLOPSALOPSIS

Meunier 1910, p. 46.

Syn: *Peridinium* Ehrenbergh partim
       Diplopsalis Bergh partim

Cell similar in shape to *Zygabikodinium* and with similar apical tabulation. The genera are separated on the basis of the number of antapical plates; *Diplopsalopsis* has two and *Zygabikodinium* one. Plate formula: 3', 2a, 7", 3c, 5"', 2"".

The cysts have a unique archeopyle formed by a suture between the second to sixth precingular plates and the two intercalary plates and the second and third apicals.

Diplopsalopsis orbicularis (Pauls.) Meun.

Fig. 18 L, M

Meunier, 1910, p. 46, pl. 3, fig. 14–17.
Lebour, 1925, p. 103–104, pl. 16, figs. 1a–e.
Wall & Dale, 1968a, p. 280, pl. 4, fig. 20, text fig. 7.

Syn: *Peridinium orbiculare* Paulsen, 1907, p. 1, fig. 10.
              Paulsen, 1908, p. 42, figs. 50a–k.
              Schiller, 1937, p. 141–142, figs. 141a–e.
       Diplopsalis orbicularis Steidinger & Williams, 1970, p. 49.

Cell similar in shape to *Diplopsalis lenticula* Bergh but can be more globular and differs in plate pattern. There is a small apical projection, both epitheca and hypotheca have convex sides. Plate 1' is 'ortho' type; the two intercalary plates are subequal but not markedly so. Girdle equatorial, not offset, excavated and bordered by lists, probably not supported by spines. The hypotheca has two antapical plates. There is a short sulcus, not reaching to centre of hypotheca and bordered on the left by a large list. Plasma pinkish. Plate formula: 3', 2a, 7", 5"', 2"".

Size: 44–53 µm long; 40–68 µm diameter.

Distribution: North Sea, English Channel. Not common. Also in Danish and Icelandic waters, the Baltic Sea and the Pacific Ocean.

Cysts: These are dark brown, weakly oblate and microgranular to scabrate. Traces of sutural tabulation may be visible on the cysts. The archeopyle is delimited by a jagged suture on the dorsal and lateral surface of the epitract and the archeopyle lid corresponds to the three apical plates and the two intercalary plates in the thecate stage.

Cyst size: 45–52 µm long; 46–62 µm diameter.

Occurrence: Rare in Pleistocene and Holocene sediments.

Some dispute over the generic position of this species exists based on the plate formula. Some workers have favoured its inclusion in *Peridinium* (now *Protoperidinium*), others in *Diplopsalis*. It is very similar to that of *Dissodium*. The resting spore archeopyle is unique amongst those that have been studied and this leads Wall & Dale (1968a) to support Lebour's (1922, 1925) view that it should be kept as a distinct genus.

DISSODIUM Abe

Abe 1941, p. 129

Syn: *Diplopelta* Jorgensen, 1913.
        Glenodinium Ehrenberg partim
        Peridiniopsis Lemmermann partim.

Cells similar to *Diplopsalis* but distinguished from this genus by having two antapicals instead of one.

Plate formula: 3', 2a, 6", 3c, 6s, 5''', 2''''.

Type species: *D. parvum* Abe.

The species included here were formerly placed in the genus *Peridiniopsis* but the work of Bourrelly (1968) has shown that the type of that genus has six girdle plates. Unlike Taylor (1976) who continues to use *Peridiniopsis* for these marine species, it seems appropriate to use Abe's *Dissodium* which has the correct plate formula. Unfortunately there is some likelihood of confusion with the rather similar name, *Dissodinium,* used for certain parasitic dinoflagellates.

Dissodium asymmetricum (Mangin) Loeblich

Fig. 18F, G

Loeblich, 1970, p. 905.
Drebes, 1974, p. 132, fig. 113.

Syn: *Diplopsalis lenticula* Stein, pl. 8, figs. 12—14, and
pl. 9, figs. 1—4.
D. lenticula f. *asymmetrica* Steidinger & Williams, 1970, p. 49, fig. 52.
Peridiniopsis asymmetrica Mangin, 1911, p. 644, figs. 1—4.
P. asymmetrica Lebour, 1922, p. 798, figs. 6—10.
Lebour, 1925, p. 101, pl. 15, figs. 3a—e.
Taylor, 1976, p. 132, pl. 28, figs. 296a, b.
Peridinium lenticula Paulsen, 1912, p. 265 in part.
Diplopelta bomba Jorgensen, 1913.
D. symmetrica Pavillard, 1913, p. 4, fig. 1.
Balech, 1964b, p. 22.
Praeperidinium asymmetricum Mangin, 1913.
Peridinium asymmetricum Ostenfeld, 1915.
Glenodinium lenticulum f. *asymmetrica* Schiller, 1937, p. 105,
figs. 97a—e.
Diplopsalis asymmetrica Drebes & Elbrachter, 1976, p. 80.

Cell lenticular to globular, circular in apical view. Epitheca with convex sides ending in a small apical projection surrounding the apical pore. First apical of 'ortho' type reaching from girdle to apex. The two intercalary plates are unequal, 1a being a very small rhomboid plate bordered by 2', 2a, 2" and 3". 2a large, dorsally situated, almost reaching the apex. There are six precingular plates. Intercalary striae may be present. The girdle is equatorial and not displaced or excavated but bordered by lists, typically without supporting spines. Hypotheca equal in depth to epitheca with convex sides. Five postcingular and two antapical plates present. The sulcus is short, reaching only three-quarters of the way down the hypotheca and with a short but relatively wide list on the left side. The cell contents are pink, a large pusule may be present and the nucleus is central. Theca finely punctate.

Size: 32—70 µm long; 60—95 µm diameter.

Distribution: A common species recorded from all round the British Isles. Also from the Baltic, Atlantic, Mediterranean, Indian Ocean and Gulf of Mexico.

Note: Frequently confused or recorded collectively with *Diplopsalis lenticula* and *Zygabikodinium lenticulum* but usually larger than those species.

Dissodium excentricum (Nie) Loeblich

Fig. 18H

Loeblich, 1970, p. 905.

Syn: *Diplopsalis excentrica* Nie, 1943, p. 17–19, figs. 32–36.
       Peridiniopsis excentrica Taylor, 1976, p. 132.

A large species characterised by the small apical plates and very large second intercalary plate. The body is slightly flattened at the anterior and posterior to give a sub-ellipsoidal or lens-shaped appearance in ventral view. There is no apical horn on the epitheca which is rounded. 1' is orthoperidinium type and is less than half the height of the epitheca. 2' and 3' are also small and extend to just over half way up the epitheca. 1a borders 2', 2" and 3" and is sub-rectangular in shape. The anterior portion of the epitheca is covered by plate 2a which occupies one half of the total epitheca. There are six precingular plates. The cingular region is bordered by narrow lists, not supported by spines; the girdle is not excavated. The girdle is made up of three plates, 1 and 3 are small and ventral, 2c forming the remainder of the girdle. There are five postcingular plates and two antapicals. There are six sulcal plates. The theca bears scattered pores. Wide intercalary striae may be present.

Plate formula: 3', 2a, 6", 3c, 6s, 5"', 2"".

   Size: 55–60 μm x 66–77 μm
           Length : breadth 0.80–0.84 : 1

Distribution: In coastal waters. Found in the North Sea off Blyth and off eastern Scotland from Fraserburgh to the Firth of Forth. Otherwise known only from Shin-tsuen-Kong (in Hainan, in the South China Sea), the type locality.

OBLEA Balech

Balech 1964b, p. 19.

Syn: *Peridiniopsis* Lemmermann partim
       Glenodinium Ehrenberg partim

Cells globular or occasionally lenticular with prominent left sulcal list. Epitheca with markedly asymmetrical tabulation; three apical plates, the first five-sided of 'meta' type and there is only one large anterior intercalary plate, otherwise the tabulation is similar to that found in *Dissodium,* i.e. 3', 1a, 6", 3c, 5"', 2"". Thecal sculpturing absent or consisting of scattered pores. As in *Dissodium* the sulcus is short, not reaching to the centre of the hypotheca.

Type species: *O. baculifera* Balech.

Oblea rotunda (Lebour) Balech ex Sournia

Fig. 18A, B

Balech 1964, p. 23.
Sournia 1973, p. 49.

Syn: *Peridiniopsis rotunda* Lebour, 1922, p. 804, figs. 16, 20.
Lebour, 1925, p. 101, pl. 15, figs. 4a–e.
Glenodinium rotundum Schiller, 1937, p. 107, figs. 98a–e.
Diplopsalis rotunda Wood, 1968, p. 54, fig. 134.
D. rotundata Steidinger & Williams, 1970, p. 49.

Cell globular with small apical prominence, central girdle and wing
bordering sulcus. Epitheca with convex sides and a 'meta' type of first apical
plate and the other two apical plates subequal in size. Anterior intercalary
plate large touching five of the six precingular plates. Girdle bordered by
prominent lists supported by spines, not excavated, and made up of three
plates. Hypotheca with convex sides and no antapical spines, consisting of
five postcingular and two antapical plates. The sulcus is bordered on the left
side by a prominent list. Plasma pinkish or colourless with a conspicuous
pusule in the hypotheca and nucleus situated in the girdle region. Plate
formula: 3', 1a, 6", 3c, 5''', 2''''.

Size: 22–34 µm diameter.

Distribution: Found all round coast of Britain from March to
November. Also recorded from Australia, Antarctic, Straits of Florida and
Gulf of Mexico.

ZYGABIKODINIUM

Loeblich & Loeblich III, 1970, p. 541

Syn: *Diplopeltopsis* Pavillard, 1913

Cells globular or lenticular with an 'ortho' type of first apical, one
or two intercalary plates and a single antapical plate. The sulcus is bordered
on the left side by a list.

Plate formula: 3', 2a, 7", 3c, 6s, 5''', 1''''.

Type species: *Z. lenticulatum* (Mangin) Loeblich & Loeblich.

Loeblich & Loeblich III renamed this genus because of the pre-
occupation of *Diplopeltopsis* by a Lichen; however, this change would not be
deemed necessary by zoologists following their International Code.

Zygabikodinium lenticulatum (Paulsen) Loeblich & Loeblich

Fig. 18C, D

Loeblich & Loeblich III, 1970, p. 541

Syn: *Diplopsalis lenticula* f. *minor* Paulsen, 1907, p. 9, fig. 9.
                                        Steidinger & Williams, 1970, p. 49, fig. 53.
        Peridinium lenticulum Mangin, 1911, p. 30, figs. 3–4.
        Diplopeltopsis minor Pavillard, 1913, p. 7.
                                        Lebour, 1922, p. 801–804, figs. 11–15.
                                        1925, p. 102, pl. 15, figs. 2a–e.
                                        Wall & Dale, 1968a, p. 208, fig. 7, pl. 4,
                                                                figs. 21, 22.
                                        Drebes, 1976, p. 131, figs. 112a–c.
        Glenodinium lenticula forma *minor* Schiller, 1937, p. 105.
        Diplopsalis minor Wood, 1968, p. 54, fig. 133.

Cell globular to lens-shaped, almost circular in apical view. Epitheca with convex sides and small apical projection. First apical plate a narrow 'ortho' type but with anterior triangle longer than posterior triangle. There are three apicals and one very small square intercalary plate on the left side and a large intercalary in a dorsal position almost meeting the apical pore. There are seven precingular plates. The girdle is equatorial, not incised but bordered by lists supported by spines. The hypotheca has convex sides composed of five postcingular plates and only one antapical. The sulcus almost reaches the centre of the hypotheca and is bordered by a narrow list on the left. Cell contents pink. Theca delicately punctate.

Size: Diameter 28–56 μm, usually over 40 μm.

Distribution: Recorded from all round the British coast, except N.W. Scotland. More abundant in the south. Also from the Baltic, Australia, Brazil and Caribbean Sea.

Note: Frequently recorded collectively with *Diplopsalis lenticula* and *Dissodium asymmetricum*.

Cysts: Described by Wall & Dale (1968a) as small, slightly oblate, cysts which appear to be enveloped in a thin, wrinkled outer membrane having a reflected tabulation, cingular ridges and a small projection at the site of the original sulcal list. The archeopyle is formed from a split between the anterior girdle border and the precingular plates; the epitractal area remains attached to the cingular and hypotractal areas both laterally and ventrally.

Cyst size: 40–60 μm wide.

Reid (1977) recorded cysts which he called *Dubridinium cavatum* and *D. caperatum* which were often enclosed in thecae of *Zygabikodinium lenticulum*. These cysts were fairly common around the coasts of the British Isles, particularly to the west and south.

Zygabikodinium pseudooblea Elbrachter

Fig. 18E

Elbrachter, 1975, p. 60, figs. 5a–d.

Small globular cell with marked apex. Similar in shape to *Oblea rotunda* but with different plate pattern. Epitheca with convex sides, 1' 'ortho' type, one intercalary plate is dorsal, the other is situated on the right side. The girdle is slightly excavated, bordered by narrow lists, its width is almost one fifth of body length. The hypotheca is convex and has a single large antapical plate. The sulcus is shallow and bordered on the left side by a large list.

There are no chromatophores but two large pusules are present in the hypocone and the elliptical nucleus is situated dorsally in the region of the cingulum. Plasma clear; nutrition may be saprophytic.

Plate formula: 3', 2a, 7", 3c, 5''', 1''''.

Size: 23–28 μm long; 23–28 μm diameter.

Distribution: Off List/Sylt in the North Sea from March to June.

SCRIPPSIELLA Balech

Balech, 1959, p. 196.

Small dinoflagellates with a conical epitheca and rounded hypotheca separated by a median excavated girdle. Chloroplasts present, nucleus ovoid, centrally placed. Thecal plates relatively thin and unsculptured.

Plate formula: 4', 3a, 7", 5–6c, 5''', 2'''', 4s.

Type species: *S. sweeneyae* Balech.

Note: Members of this genus are thecate, with tabulation differing from *Protoperidinium* and *Peridinium* in the girdle and sulcal regions only. The two species commonly found around the British Isles, *S. trochoidea* and *S. faeroense* have caused several taxonomic problems. Superficially, they are very similar but differ in size of first apical plate, number of cingular and sulcal plates and in the type of cyst produced. The present situation is very confused and here these two species have been kept together in the genus

Scrippsiella although it has recently been suggested that they should be in different genera. (Dale, 1977). *S. trochoidea* has six cingular plates and produces calcareous cysts comparable to those of *S. sweeneyae* and is therefore a 'true' *Scrippsiella* whereas *S. faeroense* has only five plates and organic walled cysts showing affinities with *Peridinium,* not however with *Protoperidinium* which has three cingular and one transitional plate. Dale's proposition that *S. faeroense* should be included in the genus *Peridinium* means that this species would currently be the only marine representative of the genus. All previously named *Peridinium* species occurring in marine situations having been transferred to *Protoperidinium* because of their reduced number of cingular plates. Work on the life cycles is needed to completely resolve the confusing state of this genus.

Scrippsiella faeroense (Pauls.) Balech & Soares

Fig. 18 N–P, Pl. III e, f.

Balech & Soares, 1967, p. 106, figs. 11–20.

Syn: *Peridinium faeroense* Paulsen, 1905, p. 5, fig. 5.
                    Lebour, 1925, p. 113, pl. 19, figs. 2a–d.
                    Schiller, 1937, p. 137, figs. 134f, g.
                    Dale, 1977, p. 241, figs. 1–5, 8, 10–11, 14–19,
                                        21, 23–25, 27–31.

Small pyriforme thecate dinoflagellate with median girdle, conical epitheca with rounded apex and a rounded hypotheca. Very similar in shape to *S. trochoidea,* indeed several authors have suggested that they may be conspecific (e.g. Taylor, 1976). Narrow 'ortho' first apical (in *S. trochoidea* a wider first apical occurs) and three intercalary plates. The girdle is wide, excavated, composed of five plates and bordered by very narrow lists. The deep and wide sulcus is restricted to the hypotheca and consists of four plates. The hypotheca is without antapical spines and the sulcus does not indent the antapex. The thecal plates have been shown to have pores surrounded by up to five concentric ridges (Dale, 1977, fig. 31). Broad intercalary bands sometimes occur. The nucleus is large, median and spherical to ovoid. Chloroplasts, described as brown by Lebour (1925) and yellowish-green by Dale, are spherical or ribbon-like. The longitudinal flagellum is approximately as long as the cell.

Plate formula: 4', 3a, 7", 5c, 5''', 2'''', 4s.

Cysts: These are spherical with a variable number of solid processes with distinctive distal terminations which may be branched or unbranched (Dale, 1977, p. 243, figs. 33–35). The cyst wall is organic and resembles sporopollenin in its resistance to acid treatments. No definite archeopyle is discernible as excystment appears to occur by a simple split in the cell wall. (For further descriptions of cysts see Dale, 1977, p. 243–254.)

Size: Motile stage : 28–35 μm long, 15–29 μm wide.
       Cysts : 19–36 μm diameter.

Distribution: Occurs all round coast of Britain. Elsewhere recorded from Norway, western North Atlantic and N.E. Pacific Ocean. Frequently recorded together with *S. trochoidea*.

Note: As mentioned earlier, this species closely resembles *S. trochoidea*. Its generic status is unclear since it has five cingular plates distinguishing it from the other *Scrippsiella* species which produce calcareous cysts. *S. trochoidea* has a more pronounced apical horn, and a wider first apical plate.

Scrippsiella trochoidea (Stein) Loeb. III

Fig. 18 Q–S, Pl. III d.

Loeblich III, 1976, p. 25.
Steidinger & Balech, 1977, p. 69–73.
Dale, 1977, p. 246, fig. 13.

Syn: *Glenodinium trochoideum* Stein, 1883, pl. 3, figs. 27–29.
                          Paulsen, 1908, p. 24.
      G. acuminatum Jorgensen, 1899, p. 32.
      Peridinium trochoideum Lemmermann, 1910, p. 336.
                          Lebour, 1925, p. 113, pl. 19, figs. 3a–d.
                          Schiller, 1937, p. 137, figs. 134a–e.
                          Wall & Dale, 1968b, p. 1401, fig. 2, 1–3,
                                      pl. 172, figs. 1–4, 27.
                          Drebes, 1973, p. 141, fig. 123c.
      Scrippsiella faeroense Dickensheets & Cox, 1971, p. 139, figs. 1–6.

Small pyriforme cell with conical epitheca. The girdle is wide, excavated, composed of six plates and surrounded by narrow lists. The epitheca has orthohexa tabulation. There are no antapical projections and the wide excavated sulcus does not indent the antapex. The theca has scattered poroids. Several yellow-brown chromatophores are present and the nucleus is central.

Plate Formula: 4', 3a, 7", 6c, 5''', 2''''.

Size: 16–36 μm long; 20–23 μm broad.

Distribution: Occurs all round coast of Britain. Elsewhere from Atlantic, West Indies, frequently recorded together with *S. faeroense*.

Cysts: Spherical, covered with more than fifty calcareous spines. The latter are simple, rodlike and blunt, pointed, or with a small distal knob. (Wall & Dale, 1968b).

Size: 28–36 μm diameter.

Distribution: Probably found in most temperate waters.

Note: This species closely resembles *S. faeroense* which is considered by some workers to be conspecific. It differs in having a wider first apical, six instead of five cingular plates and produces calcareous rather than organic walled cysts.

PROTOPERIDINIUM Bergh 1882

Syn: *Peridinium* Ehrenberg 1831 partim

Introduction

Thecate dinoflagellates with distinct epitheca and hypotheca separated by a girdle which is usually equatorial and may or may not be displaced, with or without overhang. The sulcus may slightly impinge on the epitheca. The body shape is variable. Chromatophores and chloroplasts are usually absent.

The genus was created by Bergh (1882) for the new species *Protoperidinium pellucidum,* and to include *Peridinium michaelis* Ehrenberg. Gran (1902) made the first split of the genus *Peridinium* into groups, adopting Bergh's *Protoperidinium* for species with solid antapical spines and a right-handed girdle and making the group *Euperidinium* to include those species characterised by hollow antapical horns and a left-handed girdle. Paulsen (1908) followed these divisions in his work on thirty-nine species of *Peridinium,* although some difficulty was encountered with species with characters from both divisions. Broch (1910), following the same method of grouping, pointed out that the relationship between plates was significant for identification. A treatment of the genus using thecal plate pattern was made by Jorgensen (1912). His divisions were based on the shape of the first apical plate and the position and shape of the dorsal epithecal plates particularly the second intercalary plate (2a). He did not use the classification put forward by Gran but presented new groups: Sub-genus *Archaeoperidinium* – those forms with only two anterior intercalary plates. Those with three intercalary plates he considered to be true peridinea, Subgenus *Peridinium,* and two groups were then made as follows:

Orthoperidinium – Those species with the first apical plate bordering four major epithecal plates (ortho). He divided these into three sections on the basis of their dorsal plates – *Tabulata, Conica* and *Oceanica.*

Metaperidinium – First apical with five or six sides (meta or para). These in turn divided into *Pyriformia, Paraperidinium, Humilia* and *Divergens.*

Barrows (1918), Lindemann (1918) and Lebour (1925) adopted Jorgensen's classification but noted that plate variation can occur within a species. Abe (1927) also accepted this scheme but stated that the characters of the appendages and the girdle should be considered together with the plate pattern when defining a species. In 1930 Peters elevated the division *Paraperidinium* into a subgenus, united the sections *Divergens* and *Humilia* and made two new sections, *Areolata* and *Varigata*. Paulsen (1931) adopted the Jorgensen system with certain amendments. He divided the subgenus *Archaeoperidinium* into two sections namely *Avellana* to include those species with a symmetrical epitheca, and *Excentrica* for those species with an asymmetrical epitheca. He introduced the name *Veroperidinium* to include the peridinea with three anterior intercalaries. He kept Jorgensen's sections *Pellucida, Humilia, Pyriformia, Tabulata, Divergenia, Oceanica, Conica,* and created a new section *Paradivergens.*

In 1937 Schiller's monograph of Dinoflagellates was published and included 488 species of marine and freshwater peridinia, grouped into the sections used by Paulsen. Abe (1936) published detailed descriptions of numerous peridinea including drawings of sulcal plates. Graham (1942) emphasised the importance of sulcal plates in the taxonomy of the peridinea. He also suggested a numerical method for describing species. Since then workers have used the modified Jorgensen's system of classification but have paid more attention to the details of plate morphology and sculpture. The introduction of the scanning electron microscope has been valuable in this respect. Much attention has been paid to the specific differences in sulcal plates. One notable worker in this field is Balech who in 1974, following the lead of Loeblich (1970) utilised Bergh's name *Protoperidinium* to include all the peridinea with three girdle plates, and left *Peridinium* to include only those with five or six girdle plates. In general this divides the marine species (in *Protoperidinium*) from those found in freshwater (*Peridinium*).

Thecal plates

The typical plate formula is 4', 3a, 7", 3c, 5"', 2"" (see Fig. 1 C–F; Fig 22). There may however be two or rarely four intercalary plates. There is an apical platelet surrounding the apical pore. A transition plate is situated between the girdle and the sulcus, the latter being composed of six or seven plates (Fig. 22J). The importance attached to the shape and position of these plates in relation to the taxonomy of the genus has already been mentioned and for more detailed information reference may be made to the works mentioned above. In general, however, these plates are not readily seen and they must be separated by loosening in sodium hypochlorite. This is time consuming and not practical for routine identification. For this, body shape, ornamentation and thecal plate structure are used.

Classification of the species within the genus Protoperidinium
(in the order that they are described)

Sub-genus: ARCHEOPERIDINIUM

      P. avellana

      P. denticulatum

      P. excentricum

      P. minutum

      P. thorianum

Sub-genus: MINUSCULA

      P. bipes

Sub-genus: PROTOPERIDINIUM

    Group: Orthoperidinium

        Section   Oceania

            P. depressum

            P. marielebourae

            P. oceanicum

            P. oblongum

            P. saltans

        Section: Tabulata

            P. claudicans

            P. punctulatum

        Section: Conica

            P. achromaticum

            P. conicoides

            P. conicum

            P. leonis

            P. obtusum

            P. pentagonum

            P. subinerme

Group: Metaperidinium

    Section: Humilia

        P. brevipes

        P. cerasus

        P. curtipes

        P. decipiens

        P. divergens

        P. globulum

        P. ovatum

        P. subcurvipes

    Section: Pyriformia

        P. granii

        P. mite

        P. pyriforme

        P. steinii

Group: Paraperidinium

        P. curvipes

        P. diabolum

        P. islandicum

        P. pallidum

        P. pellucidum

        P. thulesense*

(* of uncertain systematic position)

Typically the markings on the surface of the plates are not pronounced although reticulation, pores, vertical and longitudinal striae are visible with light microscopy. Boltovskoy (1973), Gocht & Netzel (1974) and Balech (1976d) have published papers using scanning electron microscopy which readily reveals the surface markings.

Ornamentation in the genus is not elaborate and is confined to the girdle, sulcus and antapical region. The girdle, which may or may not be excavated, frequently has both anterior and posterior lists with or without supporting spines. The sulcus may be bordered by lists and usually the left side has the more conspicuous list which may extend beyond the posterior end of the hypotheca. The hypotheca may be drawn out into hollow horns, or plates 1 and 2'''' may bear solid spines with or without wings.

KEY TO THE SPECIES OF PROTOPERIDINIUM

(Use in conjunction with Fig. 1 and Fig. 22 which illustrate the terms used)

1a	Two anterior intercalary plates (1a, 2a)	2
b	Three anterior intercalary plates (1a, 2a, 3a)	5
2a	Cell rounded	3
b	Cell not rounded	4
3a	Cell globular; distinct low apical horn; girdle not displaced nor excavated; theca covered in random pores	*P. minutum*
b	Cell rounded but flared at girdle; no distinct apical horn; girdle left-handed and excavated; theca strongly papillate or wavy	*P. thorianum*
c	Cell ovoid; no apical horn; girdle excavated	*P. avellana*
4a	Cell flattened at antapex; epitheca conical, almost straightsided; no distinct apical horn; intercalaries equal; surface has circular thickenings pierced by pores	*P. denticulatum*
b.	Cell irregularly flattened anterio-posteriorly; apical horn excentrically placed towards ventral side; intercalaries unequal; theca finely reticulated	*P. excentricum*
5a	Six precingular plates. Length with spines 23–35 μm	*P. bipes*
b	Seven precingular plates	6

6a	Five to six cingular plates. Freshwater species	sub-genus *Peridinium*
b	Three cingular plates. Marine	7
7a	First apical plate ortho (four sided)	8
b	First apical not ortho	20
8a	Girdle left-handed or not offset	9
b	Girdle right-handed, excavated; antapex flattened; 2a quadra or hexa; sulcus deep, widest posteriorly; theca reticulated	*P. subinerme*
9a	2a quadra (see Fig.22O)	10
b	2a penta (five sided)	14
c	2a hexa (see Fig. 22M)	15
10a	Hypotheca drawn out into a distinct apical horn; girdle not excavated; theca reticulate or punctate	11
b	Hypotheca conical and not drawn out into an apical horn; girdle excavated; theca covered in small spines; dorsal intercalary bands in straight lines	*P. marielebouriae*
11a	Antapical horns straightsided	12
b	Antapical horns curved and diverging	*P. saltans*
12a	Cell equidimensional or slightly longer than broad. Protoplasm pink	*P. depressum*
b	Cell longer than broad	13
13a	Cell > 220 μm long. Oceanic	*P. oceanicum*
b	Cell <160 μm long. Neritic	*P. oblongum*
14a	Hollow unequal antapical horns; epitheca drawn out to form an apical horn; theca reticulate	*P. claudicans*
b	No horns; epitheca conical theca punctate or spiny	*P. punctulatum*
15a	First apical symmetrical	16
b	First apical asymmetrical, upper triangle shorter than lower one, right side larger than left; apex blunt; hypotheca reticulated; epitheca covered in longitudinal striations	*P. obtusum*
16a	Sutures between apical plates and precingulars 1" 2" 6" and 7" raised	17
b	No raised sutures	19

17a   Raised sutures radial   18

  b   Raised sutures form zig-zag pattern; hypotheca with
      two hollow horns each bearing a solid spine; theca
      reticulated with fine spines at junctions of
      reticulations and pores   *P. leonis*

18a   Cell rhombic; two conical hollow antapical horns
      not bearing spines; sulcus deep and extending
      length of hypotheca   *P. conicum*

  b   Cell pentagonal; antapex flattened bearing two solid
      spines; sulcus only two-thirds length of hypotheca   *P. pentagonum*

19a   Sulcus with twist on left below girdle; cell
      contents pale yellow; length 45—71 μm   *P. conicoides*

  b   Sulcus straight; cell contents colourless;
      length 27—40 μm   *P. achromaticum*

20a   First apical meta (five sided)   21

  b   First apical not meta   32

21a   2a quadra   22

  b   2a penta   29

22a   Cell flattened anterio-posteriorly   23

  b   Cell not flattened anterio-posteriorly   25

23a   Antapical spines present   24

  b   Antapical spines absent; no prominent sulcal
      lists; epitheca rising abruptly to a low
      symmetrical apical horn   *P. decipiens*

24a   Hypotheca bears two slightly diverging solid
      spines each with two wings. Cell lenticular
      in polar view   *P. ovatum*

  b   Hypotheca bears a single curved spine on right
      antapical plate; sulcus has prominent list on
      left margin and this curves round at antapex.
      Cell globular in polar view   *P. subcurvipes*

25a   Cell globular with very small 1″ plate   26

  b   Cell not globular   27

26a	No antapical spines; sulcus narrow not reaching to centre of hypotheca	*P. globulum*
b	Two long fine antapical spines ±wings; sulcus broadens posteriorly and extends length of hypotheca	*P. cerasus*
27a	Cell with two hollow antapical horns bearing spines; sides of body concave	28
b	Cell with no antapical horns, two small antapical spines. Cell rhombic with straight-convex sides; 2a very small and almost square	*P. brevipes*
28a	Cell longer than broad; protoplasm pink	*P. divergens*
b	Cell diameter greater than length; protoplasm yellow	*P. curtipes*
29a	Cell with no hollow antapical horns but with spines	30
b	Cell with hollow antapical horns ±wingless spines; cell straight-convex sided; sulcus narrow with lists projecting antapically	*P. granii*
30a	Antapical spines bearing wings	31
b	Diverging antapical spines without wings; cell with rounded sides and pronouced apical horn; girdle not excavated	*P. mite*
31a	Girdle excavated; two antapical spines, left one in line with sulcus and bearing wing which stretches along sulcus to the girdle	*P. pyriforme*
b	Girdle not excavated; two very long antapical spines each with three wings none of which is continuous with the sulcal lists	*P. steinii*
32a	First apical para (six sided)	33
b	First apical triangular; sutures radiating from apex to girdle, plate formula difficult to interpret; epitheca conical; hypotheca with straight sides and flattened apex	*P. thulesense*
33a	Two straight, winged antapical spines and a posterior projection of the left sulcal list gives appearance of a third spine	34

33b   One straight antapical spine on the right and
      on the left the sulcal list protrudes as a
      well-developed curved wing; protoplasm pale
      yellow   *P. curvipes*

34a   Cell longer than broad; distinct apical horn   35

  b   Cell depressed from above, broader than long;
      epitheca conical; theca smooth or poroid;
      antapical spines short   *P. islandicum*

35a   Antapical spines < half length of hypotheca;
      apical horn not greatly attenuated   36

  b   Antapical spines equal to or greater than the
      length of hypotheca; apical horn greatly
      attenuated; protoplasm clear   *P. diabolum*

36a   Cell length >70 μm; robust spines and wings;
      cell contents pinkish; theca finely dotted   *P. pallidum*

  b   Cell length <70 μm; delicate spines and lists;
      chromatophores absent; theca reticulated   *P. pellucidum*

Sub-genus: Archeoperidinium

Protoperidinium avellana (Meunier) Balech

Fig. 19F–H

Balech, 1974, p. 54.

Syn: *Properidinium avellana* Meunier, 1919, p. 56, pl. 18, figs. 37–41.
Peridinium avellana Lebour, 1925, p. 108, pl. 17, figs. 1a–f.
Schiller, 1937, p. 139, figs. 137a–e.
Wall & Dale, 1968, p. 277, pl. 3, fig. 29.

The cell is ovoid, being longer than broad, and the girdle is median.
The epitheca is rounded with a long, flat, apical pore. The hypotheca is
broadly rounded and the antapex has a shallow depression caused by the
sulcus. The girdle is fairly deep and the girdle plates have vertical striations.
There are only two anterior intercalary plates, both situated on the ventral
side of the epitheca. The thecal plates are 'smooth or micro-reticulate' (Wall &
Dale) or 'thickly covered with pores and striae' (Lebour). The cell contents
are greenish yellow or colourless.

Size: 54–63 μm long; 30–58 μm wide.

Distribution: Originally described from brackish water on the coast
of Belgium. Also found in Plymouth Sound, and at Woods Hole, U.S.A.

Cysts: Found by Wall & Dale (1968) at Woods Hole, the cysts are
spherical, smooth-walled and dark brown. There is a distinct hexagonal
archeopyle which probably corresponds to plate 2a. The cysts are 44–
63 μm in diameter.

Wall & Dale state that these cysts are identical with fossil cysts called
Chytroeisphaeridia cariacoensis from Holocene sediments of the southern
Caribbean. The cysts of *Protoperidinium punctulatum* and *P. denticulatum*
are also extremely similar.

Protoperidinium denticulatum (Gran & Braarud) Balech

Fig. 19A, B

Balech, 1974, p. 54.
Schiller, 1937, p. 519, fig. 606.

Syn: *Peridinium denticulatum* Gran & Braarud, 1935, p. 381, fig 58.
P. clavus Abe, 1936, p. 661, figs. 33–51.

This species has a characteristic shape. The epitheca is conical with a
wide base and almost straight sides. There is a distinctive slit-like groove at
the apex with the apical pore at the ventral end. The species belongs in
Jorgensen's Archaeperidinium group having only two anterior intercalary

of this flat area Gran and Braarud described a row of irregular teeth. The girdle is excavated with narrow lists supported by conspicuous ribs. It is left-handed, displaced about one girdle width. The sulcus is deep. Cells can be seen in pairs attached by their hypothecae.

Plate formula: 4', 2a, 7", 5"', 2""

Size: length 32—43 μm; breadth 43—76 μm.

Distribution: Recorded from the North Sea and North of Scotland. Also from the Gulf of Maine, Helgoland and Norway.

Cysts: These are spherical, dark brown and smooth-walled and penetrated by an equatorially elongate intercalary hexagonal archeopyle (Wall & Dale 1968a, p. 277.).

Protoperidinium excentricum (Pauls.) Balech

Fig. 19J, K

Balech, 1974, p. 54.

Syn: *Peridinium excentricum* Paulsen, 1907, p. 14, figs. 17a—f.
Lebour, 1925, p. 108, pl. 18, figs. 1a—d.
Schiller, 1937, p. 144, figs. 145a—g.
Abe, 1936, p. 681, fig. 84.
Wall & Dale, 1968a, p. 278, pl. 4, fig. 8.
Faure-Fremiet, 1909, p. 228, fig. 14, pl. 16, fig. 16.

This is a distinctive species with the body irregularly flattened anterio-posteriorly and the apical horn excentrically placed towards the ventral side. The first apical plate is of the ortho type, small and connected to the apical pore by a ventral slit and the species characteristically has only two unequal intercalary plates. The hypotheca is flattened in the ventral half. The sulcus is deeply excavated and reaches beyond the centre of the hypotheca, the left side projecting further than the right. The girdle is excavated, bordered by lists supported by spines. In apical view the cell is round and the asymmetry of the cell not obvious. The protoplasm is pink. Theca finely reticulated.

Size: 36—48 μm long; 40—84 μm wide.

Distribution: Found all around the coasts of Britain; believed to be neritic. Also recorded from the Mediterranean, Indian Ocean, Pacific and Straits of Florida.

Cysts: Wall & Dale (1968a) found dark brown flattened cysts inside thecae of this species; the shape of the archeopyle is unknown.

Protoperidinium minutum (Kof.) Loeblich II

Fig. 19C, D, Pl. IV b, c

Loeblich III, 1969, p. 905.
Fukuyo, Kittaka & Hirano, 1977, p. 11, figs. 2—14.

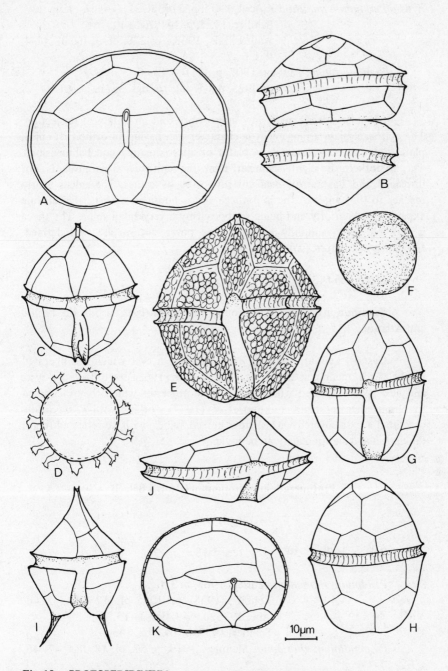

Fig. 19   PROTOPERIDINIUM

A. *P. denticulatum* apical view   B. *P. denticulatum* pair after division
C. *P. minutum*   D. *P. minutum* cyst (after Wall & Dale)   E. *P. thorianum*
F. *P. avellana* cyst (after Wall & Dale)   G. *P. avellana* ventral view
H. *P. avellana* dorsal view   I. *P. bipes*   J. *P. excentricum* ventral view
K. *P. excentricum* apical view

Syn: *Peridinium minutum* Kofoid, 1907, p. 310, pl. 31, figs. 42—45.
Schiller, 1937, p. 141, fig. 140.
Wall & Dale, 1968a, p. 278, pl. 3, fig. 32, pl. 4, figs. 6—7.
P. monospinum Paulsen, 1907, p. 12, figs. lla—d
Lebour, 1925, p. lO7. pl. 16, figs. 3a—h.

Globular species with low apical horn and an antapical projection. The epithecal plate arrangement is characteristic, having an ortho first apical plate and only two intercalary plates of approximately equal dimensions. The hypotheca is slightly smaller than the epitheca. The girdle is not displaced and has narrow lists not supported by spines. The sulcus nearly reaches to the centre of the hypotheca and is bordered by lists, the left one expanding posteriorly and being supported by a very small spine. The theca is covered with randomly scattered fine pores; sutures are well defined, frequently with intercalary striae.

Size: 23—43 µm long; 23—56 µm wide.

Distribution: Found all around the British Isles, also in the Pacific and Atlantic.

Cysts: Lebour, 1925, p. 107, recorded a cyst which was liberated by the theca opening at the girdle. Wall & Dale (1968a) record similar cysts and also spores covered with short hollow processes which they incubated to obtain thecate stages. Fukuyo *et al.* (1977) working with *P. minutum* described a cyst covered with slender curved spines, with no reflected thecal tabulation and an intercalary archeopyle.

Protoperidinium thorianum (Pauls.) Balech

Fig. 19E

Balech, 1973, p. 347, pl. 1, figs. 1-18.

Syn: *Peridinium thorianum* Paulsen, 1905, p. 3, fig. 1.
Lebour, 1925, p. 108, pl. 17, figs. 2a—d.
Abe, 1936, p. 649, figs. 1—15.
Schiller, 1937, p. 142, fig. 143.
Properidinium thorianum Meunier, 1919, p. 57, pl. 18, figs. 42—46.

Rounded cell characterised by strong papillate or wavy thecal surface and presence of only two epithecal intercalary plates. In ventral view the sides are rounded. The first apical plate is ortho with apical pore plate long, extending onto third apical plate. Sulcus straight, narrow and not usually reaching to centre of hypotheca; there may be short delicate spines at the bottom of the sulcus. Girdle left-handed, striated, excavated and bordered by small lists not supported by spines. Cell contents yellow.

Size- 56–85 μm long; 56–85 μm wide.

Distribution: Occurs sporadically all around coast of Britain. Also recorded from Iceland, Faroes, Skagerak, coast of Norway, Canada, California, Okhotsk Sea and coast of Hokkaido, and Indian Ocean.

Sub-genus: Minisculum

Protoperidinium bipes (Pauls.) Balech

Fig. 19 I

Balech, 1974, p. 53.

Syn: *Glenodinium bipes* Paulsen, 1904, p. 21, figs. 3, 4..
        Peridinium minisculum Pavillard, 1905, p. 57, pl. 3, figs. 7–9.
                        Schiller, 1937, p. 194, fig. 190.
        Miniscula bipes Lebour, 1925, p. 138, pl. 29, figs. 3a–b.

Belonging to Balech's (1974) subgenus *Minisculum* this species has six precingular plates. Cell flattened dorsoventrally with triangular epitheca ending in a long apical horn and a shorter hypotheca ending in two fine, solid, antapical horns. The thecal plates are delicate. The first apical is of meta type and 2a is penta. In ventral view, the sides of the epitheca may be straight or concave and the hypotheca straight-sided with the posterior end flattened. The antapical spines are situated at the angle between the flattened antapex and the sides of the hypotheca and diverge outwards. The girdle is slightly right-handed, excavated and bordered by lists. The sulcus widens posteriorly. The cell contents are colourless which, together with the relatively small size, may cause it to be overlooked.

Plate pattern: 4', 3a, 6", 5''', 2''''.

Size: 20–35 μm long; 17–19 μm wide.

Distribution: Neritic; found all around Britain Greenland, Iceland, Baltic, Bosporus and Mediterranean.

Note: Because of their delicate nature the thecal plates in this species are difficult to observe. The published plate patterns do not look convincing and more work is required here.

Sub-genus: Protoperidinium. Group: Orthoperidinium, section Oceania.

Protoperidinium depressum (Bail.) Balech

Fig. 20A, Pl. V e

Balech, 1974, p. 57.

Syn: *Peridinium depressum* Bailey, 1855, p. 12, figs. 33–34.
　　　　　　　Lebour, 1925, p. 119, pl. 23, figs. a–f.
　　　　　　　Schiller, 1937, p. 250, fig. 251.
　　　　　　　Graham, 1942, p. 99.
　　　　　　　Drebes, 1974, p. 137, figs. 119a–b.
　　　　　　　Gocht & Netzel, 1974, p. 381, pl. 43–52.

　　　Cell flattened obliquely in a dorsoventral plane. Prominent apical horn and two long, hollow antapical horns present. Ortho quadra plate arrangement. Theca reticulated. The antapical horns each have an internal projection which is a continuation of the sulcal list. The sulcus is deeply excavated. Girdle left-handed, unexcavated, bordered by wide lists supported by spines. Intercalary striae when present may be broad. Sides of the theca concave towards apex and antapex and convex near the girdle. Cell contents pink. A luminescent species.

Size: 116–200 μm long; 116–144 μm wide.

Distribution: Occurs all around the British Isles. Widespread distribution from Arctic to Antarctic waters. May be abundant at times.

Protoperidinium marielebouriae (Pauls.) Balech

Fig. 20H–J

Balech, 1974, p. 57.

Syn: *Peridinium marielebouriae* Paulsen, 1930, p. 69, fig. 40.
　　　　　　　Elbrachter, 1975, p. 60, fig. 3.
　　　P. obtusum (Karsten) Lebour, 1925, p. 121, pl. 24, figs. 2a–d.
　　　　　　　Schiller, 1929, p. 400, figs. 12a–d.
non *P. obtusum* (Karsten) Faure-Fremiet, 1980, p. 223, fig. 9.
non *P. divergens obtusum* Karsten, 1906, pl. 15, fig. 8.

　　　Cell rhombic with two hollow horns at antapex. The cell is flattened dorsoventrally. The epitheca has straight to weakly-convex sides. It has a regular ortho first apical plate and quadra 2a. On the dorsal surface the intercalary bands between the plates and the girdle are in a straight line; 2a which is trapezoid therefore touches 1a, 4" and 3a. At the end of the hollow horns there are small spines. The girdle is excavated and bordered by lists supported by ribs. The sulcus does not widen posteriorly. The cell surface is covered with small spines although rarely these are not present and fine reticulation or punctation occurs. Details of the sulcal plate arrangement have been described by Elbrachter (1975).

Size: Length 48–75 μm Breadth 45–70 μm

Distribution: Found all around the coast of Britain except the NW of Scotland. Also recorded as common in the North Sea (Elbrachter).

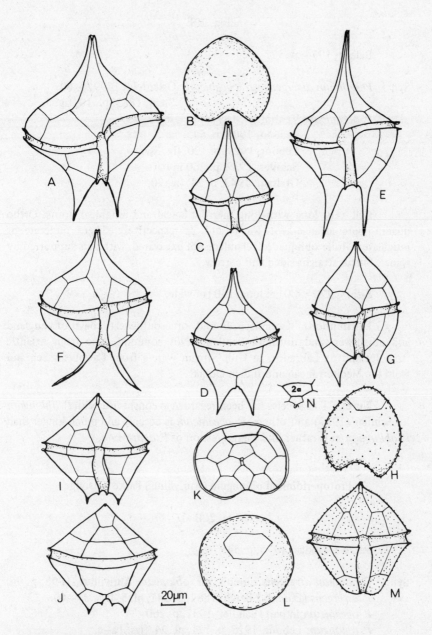

Fig. 20 PROTOPERIDINIUM
A *P. depressum* B. *P. oblongum* cyst C. *P. oblongum* ventral view
D. *P. oblongum* dorsal view E. *P. oceanicum* F. *P. saltans* G. *P. claudicans*
H. *P. claudicans* cyst and archeopyle (after Wall & Dale) I. *P. marielebouriae*
ventral view J. *P. marielebouriae* dorsal view K. *P. marielebouriae* apical
view L. *P. punctulatum* cyst (after Wall & Dale) M. *P. punctulatum* ventral
view N. 2a plate in *P. claudicans* and *P. punctulatum*

Protoperidinium oceanicum (Vanhoffen) Balech

Fig. 20E

Balech, 1974, p. 57.

Syn: *Peridinium divergens* var. *oceanicum* Ostenfeld, 1899, p. 60.
                                          Stein, 1883. pl. 10, fig. 7.
      P. oceanicum Vanhoffen, 1897, pl. 5, fig. 2.
               Paulsen, 1908. p. 54.
               Lebour, 1925, p. 120, fig. 36b.
               Schiller, 1937, p. 260 in part.
               Graham, 1942, p. 24, fig. 30.

Cell very long with long, narrow apical and antapical horns. Ortho quadra plate arrangement for 1' and 2a respectively. Theca punctate or reticulate. Girdle oblique, left-handed, not excavated, with lists supported by spines. Sulcus straight-sided and narrow.

Size: 220–300 µm long; 150 µm wide.

Distribution: Oceanic. Found off south-west coast of England and north-west and northern Scotland. Not common. Also from Atlantic from Iceland to Labrador, in Gulf Stream waters from Caribbean Sea and from the Mediterranean and Indian Oceans.

Note: This species has been regarded as conspecific with *P. oblongum* by Schiller (1937) and others. *P. oceanicum* is oceanic and much longer than *P. oblongum*, it is rather like a large version of *P. depressum*.

Protoperidinium oblongum (Aurivillius) Parke & Dodge

Fig. 20B–D

Parke & Dodge, 1976, p. 545.

Syn: *Peridinium divergens* Ehrenb. var. *oblongum* Aurivillius, 1898, p. 96.
      P. divergens (Ehrenb.) Bergh, 1882, p. 241, pl. 5, figs. 39–40.
      P. oceanicum in part, Schiller, 1937, p. 260.
      P. oblongum Lebour, 1925, p. 121, pl. 24, figs. 1a–c.
               Wall & Dale, 1968a, p. 272, pl. 1, figs. 22–29, pl. 3,
               figs. 9–11.
      Protoperidinium oceanicum Balech, 1974, in part.

Cell compressed dorsoventrally, longer than broad in dorsoventral view with long apical horn and two long, subequal antapical horns. Ortho-quadra or hexa plate arrangement. Theca finely reticulated with pores. Sulcus deep and bordered by lists. Girdle not excavated but bordered by narrow lists supported by spines. Intercalary striae when present may be very wide.

Size: Length 75–160 μm. Breadth 60–65 μm.

Distribution: Occurs sporadically all around Britain from Spring to Autumn.

Note: *P. oblongum* has been considered by many authors as a smaller neritic variety of *P. oceanicum.* Wall & Dale (1968a) confirmed the observations of Dangeard (1927) in finding a difference in the dorsal epithecal tabulation in *P. oblongum.*

Cysts: Described by Wall & Dale (1968a) as rhomboidal to cordate, dorsoventrally compressed and with a deep sulcus. Three types may be distinguished: a) Apical and antapical horns relatively long and narrow. Length:breadth ratio 1.5:1. **Size:** 74–102 x 53–77 μm. b) Sharply rounded apices and a rhomboidal outline in dorsal view. Length:breadth ratio 1.2:1. **Size:** 62–84 μm x 56–74 μm. c) Apices broadly rounded so has a cordate outline in dorsal view. Length:breadth ratio 1:1. **Size:**   47–68 x 52–68 μm.

Numerous small plastids (?) and conspicuous oil golbules, sometimes coloured bright crimson, may be present in the microgranular protoplasm. The intercalary archeopyle is hexagonal and is equivalent to 1a or 2a.

Fossil record: Some specimens are found in superficial deposits in the Arabian Sea, Atlantic Ocean and Wood's Hole. The empty resting spores are very similar to those of *Deflandrea* in size, shape and archeopyle structure but lack the thick-walled inner capsule commonly found in most Tertiary specimens.

Protoperidinium saltans (Meun.) Balech

Fig. 20F

Balech, 1973, p. 351, pl. 1, fig. 19, pl. 2, figs. 20–39, pl. 3, figs. 40–49.

Syn: *Peridinium saltans* Meunier, 1910, p. 26, pl. 1, figs. 9–14.
          Schiller, 1937, p. 263, fig. 259.

Similar to *P. depressum* but with narrower curved, diverging, antapical horns and overall smaller size. Ortho quadra plate arrangement or rarely with only two intercalary plates. Epitheca convex at girdle becoming concave distally to produce a long apical horn. Girdle not excavated or only slightly so, with lists. Details of the sulcal plates are given by Balech (1973b) who describes the sulcus as very characteristic. Theca with fine reticulations.

Size: 97–127 μm long; 66–92 μm wide.

Distribution: Occurs in North Atlantic Drift Waters off the west and north of Britain; in the North Sea in spring and summer. Also recorded from the Barents Sea and Atlantic.

Note: Some authors have considered this species conspecific with *P. depressum*. See Balech (1973b) for arguments against this.

Sub-genus: Protoperidinium. Group: Orthoperidinium, section Tabulata.

Protoperidinium claudicans (Pauls.) Balech

Fig. 20G, H

Balech, 1974, p. 57.

Syn: *Peridinium claudicans* Paulsen, 1907, p. 16, figs. 22a–d.
                               Lebour, 1925, p. 123, pl. 25, fig. 1.
                               Schiller, 1937, p. 249, fig. 250.
                               Wall & Dale, 1968a, p. 273, pl. 1, fig. 30, pl. 2,
                               figs. 1-2, pl. 3, fig. 12.

Cell dorsoventrally compressed. Epitheca drawn out into apical horn. Plates ortho and penta or quadra. Hypotheca bears two unequal hollow antapical horns (the right side longer than the left). The girdle is left-handed, excavated and bordered by lists supported by ribs. Theca finely reticulated. Said to be luminescent. Cell contents pale yellow.

Size: 50–105 µm long; 48–76 µm wide.

Distribution: Neritic; found all around coast of Britain from Spring to Autumn. Also found in coastal waters of Europe, N. America and Indian Ocean.

Cysts: Described by Wall & Dale (1968a) as cordate to pyriform in dorsal view, compressed dorsoventrally. Deep sulcal region which separates two rounded antapical lobes, of which the right is larger than the left. Surface covered by numerous short pointed spines. Archeopyle subrectangular or subpentagonal, situated subapically on the dorsal side and is equivalent to the plate 2a of the theca.

Cyst size: 47–76 µm long; 47–76 µm wide.

Cysts of *P. claudicans* are unknown in the fossil record.

Protoperidinium punctulatum (Pauls.) Balech

Fig. 20L, M

Balech, 1974, p. 58.

Syn: *Peridinium punctulatum* Paulsen, 1907, p. 19, fig. 28.
                                              1908, p. 61, fig. 79.
                            Lebour, 1925, p. 123, fig. 37.
                            Wall & Dale, 1968a, p. 276.
        P. subinerme Paulsen var. *punctulatum* (Paulsen) Schiller, 1937,
                                              p. 245, fig. 245.

Shape of cell similar to *P. subinerme* but the surface is punctate or spiny without reticulations and there are no antapical spines at the sides of the sulcus. The cell has a straight or slightly convex-sided epitheca, when seen in dorsoventral view, forming a cone. The hypotheca has concave or straight sides with a flattened antapex. The girdle is equatorial, excavated, and has lists supported by spines. The sulcus is narrow (usually narrower than in *P. subinerme*) and reaches the centre of the hypotheca. The first apical is ortho and the species was thought by Lebour (1925) to belong to the section tabulata (i.e. 2a touches 4" and 5" or 3" and 4"); this arrangement is figured by Dangeard (1927 p 354 fig 20) although there is some dispute about this and Wall & Dale (1968a, p. 276, fig. 24) show it as being in the conica section (i.e. 2a touching 3", 4" and 5").

Plate formula: 4', 3a, 7", 3c, 5"', 2"".

Size: Length 40–72 μm

Distribution: Oceanic and neritic. Found frequently in the southern North Sea but does occur all around British Isles from Spring to Autumn.

Cysts: The cysts of this species are described by Wall & Dale (1968a, p 276) as brown spherical bodies with a relatively large archeopyle (2a), hexagonal in shape. Diameter probably from 52–65 μm, archeopyle 20 x 30 μm. These cysts are difficult to distinguish from those of *P. avellana* and *P. denticulatum.*

Sub-genus: Protoperidinium. Group: Orthoperidinium, section Oceania.

Protoperidinium achromaticum (Lev.) Balech

Fig. 21B

Balech, 1974, p. 56.

Syn: *Peridinium achromaticum* Levander, 1902, p. 49, figs. 1—2.
                            Lebour, 1925, p. 114, pl. 22, figs. 1a—g.
                            Schiller, 1937, p. 229, fig. 225.

Cell rhombic in ventral view with straight or slightly convex sides.
The first apical plate is ortho and usually symmetrical. The second anterior
intercalary plate is six-sided, symmetrical and touches precingulars 3", 4"
and 5". The hypotheca is divided from the epitheca by a median girdle which
is not offset. The sulcus broadens slightly posteriorly and is bordered by
narrow lists which protrude posteriorly giving the appearance of small spines.
The girdle is slightly excavated, bordered by narrow lists. Plate formula 4',
3a, 7", 3c, 5"', 2"".

Size: Length 27—40 μm. Breadth 23—40 μm

Distribution: Neritic, brackish and marine. Has been found around
the coast of Britain though less commonly in the west. Also recorded from
Finland, Aral Sea, Colombo, West African coastal banks, Newfoundland
banks.

Note: This species can easily be confused with *P. brevipes* from
which it can be distinguished by examination of 1' which is meta in *P. brevipes*.
P. americanum (Gran & Braarud) Balech (1974) is similar in size 35—42 μm
long but has an irregular outline in ventral view and characteristically four
epithecal intercalary plates. Recorded off SW Ireland by Gaarder (1954) from
the samples taken on the 'Michael Sars' expedition in 1910.

Protoperidinium conicoides (Pauls.) Balech

Fig. 21A

Balech, 1973, p. 356, pl. 3, figs. 50—56.

Syn: *Peridinium conicoides* Paulsen, 1905, p. 3, fig. 2.
                            Lebour, 1925, p. 112, pl. 20, figs. 2a—d.
                            Schiller, 1937, p. 231, fig. 228.
                            Wall & Dale, 1968a, p. 277, pl. 2, figs. 28—30.
                            pl. 3, figs. 26—28.

Cell rhombic in dorsoventral view, nearly circular in apical view. The
conical epitheca may have slightly convex sides ending at the apex with a
small projection. Ortho-hexa plate arrangement. Hypotheca also with convex
sides ending with two hollow spines at the antapex. Girdle excavated with
fine lists which Balech (1973b) says may be absent. Sulcus deeply excavated,
broadening posteriorly, with a characteristic twist on the left just below the
girdle (plate Ss). Thecal sculpturing is delicate with random pores, intercalary
striae may occur. Cell contents pale yellow. Plate formula typical for
Protoperidinium 4', 3a, 7", 3c, 5"', 2"".

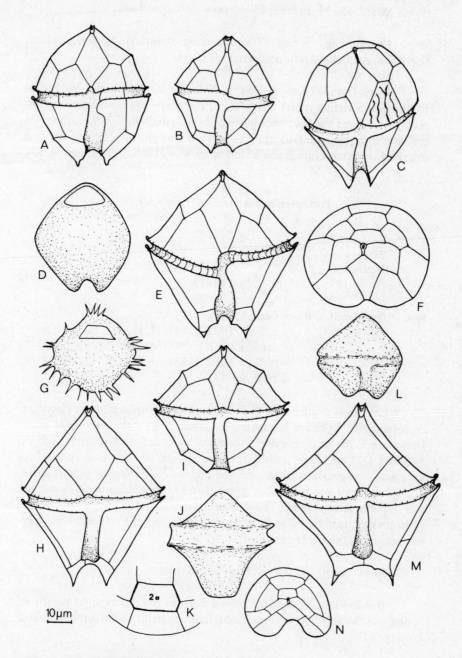

Fig. 21 PROTOPERIDINIUM
A. *P. conicoides* B. *P. achromaticum* C. *P. obtusum* (after Elbrachter 1975)
D. *P. leonis* cyst E. *P. leonis* ventral view F. *P. leonis* apical view
G. *P. conicum* cyst H. *P. conicum* ventral view I. *P. subinerme* J. *P. subinerme*
cyst K. *P. subinerme* 2a plate L. *P. pentagonum* cyst M. *P. pentagonum*
ventral view N. *P. pentagonum* apical view
(Cyst drawings are not to scale)

Size: 45–71 µm long; 45–60 µm wide.

Distribution: Found all around coast of Britain; Norway, Iceland, Greenland, Australia, Arctic and Antarctic waters.

Cysts: Wall & Dale (1968a, p. 277) recorded numerous thecae containing brown resting spores nearly circular in shape but compressed ventro-posteriorly. These have not been incubated successfully though the archeopyle has been seen by acetolysis and the spore is thought to resemble the cyst described as *Chytroeisphaeridia simplicia* by Wall (1965, p. 308, figs. 7, 20).

Protoperidinium conicum (Gran) Balech

Fig. 21G, H

Balech, 1974, p. 58
   1975, p. 21, pl. 2, figs. 12–13.

Syn: *Peridinium conicum* Gran, 1902, p. 189, fig. 14.
   Lebour, 1925, p. 111, pl. 19, figs. 1a–d.
   Schiller 1937, p. 233, fig. 229.
   Wall & Dale, 1968a, p. 273, pl. 2, figs. 3–5,
   pl. 3, fig. 13.

Cell nearly symmetrical and square when seen in dorsal view. Flattened dorsoventrally. Epitheca triangular in dorsal view with ortho-hexa plate arrangement. The sutures between the apical plates and precingulars 1–2, 2–3, 5–6 and 6–7 are characteristically raised and radial. This is in contrast to the zigzag arrangement in *P. leonis*. The deep sulcus divides the posterior end of the hypotheca into two hollow horns. Girdle excavated, bordered by narrow lists supported by spines. Theca finely reticulated and intercalary striae, when present, may be wide. Cell contents pink. Found by Tett (1971) to be luminescent in the Clyde estuary.

Size: length 70–85 µm; breadth c. 60 µm.

Distribution: Found all around coast of Britain, most frequently in Spring and Summer. Also from Mediterranean, Baltic, S. Atlantic, Amazon Estuary.

Cysts: The cysts have been described by Wall & Dale (1968a) as flattened anterioposteriorly, reniform with a very shallow sulcus when seen in polar view. Apical horn small and antapical horns absent. Cyst wall smooth and ornamented by several parallel rows of moderately long spines. The archeopyle corresponds to plate 2a. Cell contents pale and contain numerous small 'plastids'. Cysts of this type have been found in Holocene sediments in the Caribbean Sea, from the western Arabian Sea, Wood's Hole and Jamaica.

Protoperidinium leonis (Pav.) Balech

Fig. 21D–F, Pl. V d.

Balech, 1974, p. 58.

Syn: *Peridinium leonis* Pavillard, 1916, p. 32, fig. 6.
Lebour, 1925, p. 112, pl. 21, figs, la–d.
Schiller, 1937, p. 236, fig. 236.
Wall & Dale, 1968a, p. 276, pl. 2, figs. 18–21, pl. 3, fig. 22.
Balech, 1976, p. 30, figs. 2A–H.

Cell slightly flattened dorsoventrally. Epitheca and hypotheca more or less straight-sided or slightly concave. Ortho-hexa plate arrangement, rarely a fourth intercalary plate may be present but normally there are only three. Differs from *P. conicum* in having zigzag sutures from apex to girdle. Hypotheca ends in two hollow horns each bearing a single solid spine. Sulcus deeply excavated, with small lists. Girdle excavated with lists supported by spines, circular or slightly left-handed. Theca with scattered pores, reticulated with fine spines at junctions of reticulations or may bear longitudinal ridges. Luminescence of this species has been observed at Millport.

Size: Length 53–95 μm; breadth 53-95 μm.

Distribution: Occurs all around Britain, mostly from Spring to Autumn but can occur all year round. Also from Atlantic, Pacific and Indian Oceans, the Mediterranean and Caribbean.

Cysts: The cysts have been described by Wall & Dale (1968a) as depressed dorsoventrally and weakly concave in the anterioventral region. Archeopyle 2a. Two small antapical horns present with pointed tips. Deep sulcus. Wall pale-brown and either smooth or longitudinally striated.

Size: Length 52-72 μm; breadth 53–76 μm.

Protoperidinium obtusum (Karsten) Parke & Dodge

Fig. 21C

Parke & Dodge in Parke & Dixon, 1976, pp 545, 549

Syn: *Peridinium obtusum* Schiller, 1937, p. 240, figs. 241a–b.
Balech, 1949, p. 396, figs. 41–68
Hermosilla, 1973, p. 33, pl. 15, figs. 1–22.
Elbrachter, 1975, p. 61, fig. 4.
P. divergens obtusum Karsten, 1906, p. 149, pl. 23, fig. 12.
non *P. obtusum* Faure-Fremiet, 1908.
Lebour, 1925, p. 121, pl. 24, fig. 2.

This species has a blunt apex, a rounded dorsal surface and is dorso-ventrally flattened. The first apical plate is ortho but the upper (apical) triangle is much shorter than the lower (antapical) and the right larger than the left one. The hexagonal second intercalary plate touches 3", 4", 5" as well as 1a and 3a. The hypotheca has two hollow antapical horns. The girdle is inclined and the sulcus straight-sided. An unusual thecal sculpturing has been noted for this species: the epitheca is covered by longitudinal striations whereas the hypotheca is reticulated. The plasma is pinkish.

Size: Length 62 µm; breadth 53 µm.

Distribution: Found in North Sea and off Plymouth.

Note: This species has been confused with *P. marielebouriae* and *P. leonis.*. It may be distinguished from both of these by its irregular first apical plate, the inclined girdle, the surface sculpturing and the sulcal plates. Details of the latter may be found in Balech (1949), Hermosilla (1973) and Elbrachter (1975).

Protoperidinium pentagonum (Gran) Balech)

Fig. 21L–N

Balech, 1974, p. 59

Syn: *Peridinium pentagonum* Gran, 1902, p. 185, 190, fig. 15.
                          Paulsen, 1908, p. 59, fig. 76.
                          Lebour, 1925, p. 112, pl. 20, figs. la–e.
                          Schiller, 1937, p. 241, fig. 242.
       P. sinuosum Lemmermann, 1905, p. 32.
       ? *P. divergens pentagonum* Karsten, 1906, pl. 23, figs. 11a, b.
       ? *P. divergens* var. *sinuosum* Lemmermann, 1899, p. 368.

Cell pentagonal in dorsoventral view and concave ventrally. This species is characterised by the flattened antapex, the slightly displaced girdle, the short sulcus and the distinct ridges situated between the apex and the girdle, bordering plates 1', 2', 4', and 1", 2", 6" and 7". (The latter is a feature shared with *P. conicum*). The first apical plate is of ortho type; the second anterior intercalary plate is hexa, touching 3", 4" and 5". There are two solid spines antapically. Usually the girdle diameter is greater than the length of the cell. The girdle is excavated, bordered by lists and is left-handed. The nucleus is central and in the living cell the contents are pink.

Plate formula: 4', 3a, 7", 3c, 5"', 2"".

Size: Length 75–110 µm; breadth 75–100 µm.

Distribution: A neritic species found all around the British Isles. Abe (1927) recorded it off Japan; also described from Trinidad and from the Straits of Florida.

Cysts: Wall & Dale (1968a, p. 274) described the resting spores as pentagonal with broadly rounded antapical protuberances separated by a shallow sulcus. The apex is blunt in dorsoventral view and the cell slightly concave in this view. Length: breadth ratio 1.5 : 1. The cysts are colourless and the wall covered with randomly-arranged spinules. The archeopyle is a subtriangular dorsal intercalary (2a) opening.

Protoperidinium subinerme (Pauls.) Loeblich III

Fig. 21I–K, Pl. V f.

Loeblich III, 1969, p. 905.

Syn: *Peridinium subinermis* Paulsen, 1904, p. 24, figs. 10a–d.
       P. subinerme Paulsen, 1908, p. 61, fig. 79.
                    Lebour, 1925, p. 114, pl. 22, figs. 2a–f.
                    Schiller, 1937, p. 243, fig. 244.
                    Wall & Dale, 1968a, p. 276, pl. 2, figs. 22–24, pl.3, fig. 23.

Dorsal view of cell approximately square with flat base to hypotheca; epitheca with slightly convex sides and hypotheca straight-sided in this view. Ortho, quadra-hexa plate arrangement. The girdle is equatorial or slightly right-handed, excavated and surrounded by lists supported by spines. The sulcus is deep, widening antapically and having small spines at its posterior margins. These spines may be missing. The theca is usually reticulated; intercalary striae when present may be broad. Bioluminescence has been reported.

Size: Length 50–75 μm; breadth 50–60 μm

Distribution: Occurs all around the coast of Britain, also in waters off Iceland, Greenland, and in the Barents and Kara Seas.

Cysts: Described by Wall & Dale (1968a) as being relatively wide with a deeply excavated girdle zone. In polar view there is slight dorsoventral compression. Epitract concave and conical, apart from a slight convexity near the apex on the dorsal side. Small hexagonal archeopyle. Test pale-brown and smooth-walled.

Size: 36–50 μm long 48–60 μm diameter.

Distribution: Found in Caribbean and Arabian Sea and Atlantic Ocean.

Note: The motile stages of *P. subinerme* and *P. punctulatum* are similar but surface sculpturing is different (reticulated in former and punctate or spiny in latter) and sulcal region is narrower in *P. punctulatum.* The cysts are however markedly different.

Sub-genus: Protoperidinium. Group: Metaperidinium, section Humilia.

Protoperidinium brevipes (Pauls.) Balech

Fig. 22B, Pl. IV d.

Balech, 1974, p. 60.

Syn: *Peridinium brevipes* Paulsen, 1908, p. 108, fig. 151.
                        Lebour, 1925, p. 131, pl. 27, fig. 2
                        Schiller 1937, p. 200, fig. 195.
                        Balech, 1971, p. 144, pl. 30, figs. 561–577.
    P. varicans   Paulsen, 1911, p. 312, fig. 12.
                        Lebour, 1925, p. 132, fig. 416.
    P. incurvum   Lindemann, 1924, p. 2, pl. 1, figs. 7–9.

Cells rhombic with convex sides. The first apical is meta; the second intercalary plate is quadra, characteristically small and almost square, bordering plates 1a, 4" and 3a. The epitheca is slightly larger than the hypotheca. There are two small spines situated in the antapical plates either side of the sulcus. The girdle is deeply excavated and right-handed. The sulcus widens posteriorly. The protoplasm is colourless and the nucleus is large and central. Surface sculpturing is finely punctate or reticulate. Details of the sulcal plates are given by Balech (1971).

Size: Length 18–38 μm; breadth 28–34 μm.

Distribution: Neritic; found all around the British Isles in summer. Also recorded off Norway, European-African coastal banks, Argentina, Australia, Antarctic waters, Florida Straits and Brazil.

Protoperidinium cerasus (Pauls.) Balech

Fig. 22C, Pl. IV a.

Balech, 1973, p. 357, pl. 3, figs. 57–62, pl. 4, figs. 63–72.

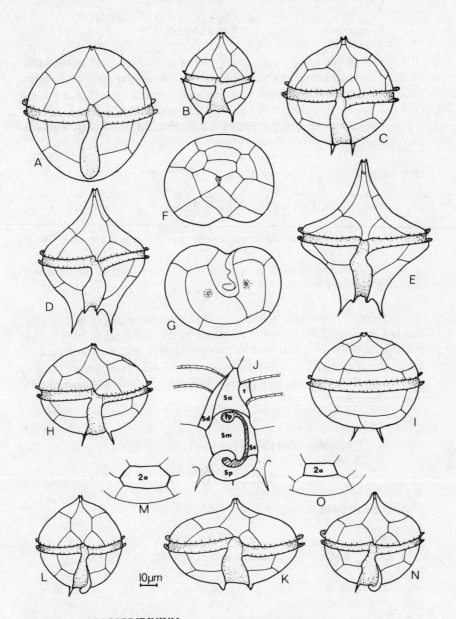

Fig. 22 PROTOPERIDINIUM
A. *P. globulum* B. *P. brevipes* C. *P. cerasus* D. *P. divergens* E. *P. curtipes*
F. *P. curtipes* apical view G. *P. curtipes* antapical view H. *P. ovatum* ventral
view I. *P. ovatum* dorsal view J. *P. ovatum* sulcal area. The Sulcal Plates are:
Sa – anterior, Sd – right, Sm – median, Sp – posterior, Ss – left, T – Transitional
K. *P. decipiens* L. *P. curvipes* M. *P. curvipes* plate 2a, hexa type N. *P. sub-
curvipes* O. *P. subcurvipes* plate 2a, quadra type.

Syn: *Peridinium cerasus* Paulsen, 1907, pl. 12, fig. 12.
                              Lebour, 1925, p. 130, pl. 27, figs. 1a—e.
       P. quarnerense Schiller, 1937, p. 184 partim.

Cell globular with small apical horn and two antapical spines. Meta-quadra plate arrangement on the epitheca; first precingular plate on the left is small. The antapical spines are long and fine, sometimes with small wings. Girdle is not excavated but has slight displacement and has lists supported by spines. The sulcus broadens posteriorly and extends slightly onto the epitheca. The plate sculpturing is finely reticulated.

Plate formula typical for genus: 4', 3a, 7", 3c, 5"', 2"".

Size: Length 46 μm; breadth 30—40 μm.

(Balech's *P. cerasus* is generally larger 53—80 μm + 6—9 μm for spines).

Distribution: Found all around British Isles. Also in the Caribbean and off S. America.

Note: There is much confusion surrounding this species, *P. quarnerense* and *P. ovatum*. For details of discussion see Schiller (1937) and Balech (1973). *P. quarnerense* however has a different tabulation of meta-hexa, a greater girdle displacement and no apical horn. Balech (1976d) has described a number of new species from the antarctic with superficially strongly resemble *P. cerasus*.

Protoperidinium curtipes (Jorg.) Balech

Fig. 22E—G, Pl. IV a.

Balech, 1974, p. 60.

Syn: *Peridinium curtipes* Jorgensen, 1912, p. 8.
                              Lebour, 1925, p. 128, fig. 39.
                              Drebes, 1974, p. 136, figs. 118a—b.
                              Elbrachter, 1975, p. 59, fig. 2.

       P. crassipes   Paulsen, 1907, p. 17, fig. 24d.
                              1908, p. 57, fig. 73.
non: *P. crassipes*   Paulsen, 1930, p. 65, fig. 36.
                              Schiller, 1937, p. 223, fig. 220, partim

Cell of slightly greater breadth than length, not dorsoventrally flattened. Walls of epitheca and hypotheca markedly concave. Thecal plates meta-quadra and strongly reticulated with spines at the junction of the reticulations. Broad intercalary striae frequent, Hypotheca ending in two hollow horns which characteristically bear projections inside, which are continuations of sulcal lists and end in spines. Girdle not displaced, deeply excavated and bordered by lists supported by spines. Cell contents yellow, not pink as in *P. crassipes*.

Size: Length 80–100 µm; breadth 67–100 µm.

Distribution: Oceanic; found all around coast of Britain. Because of the confusion between this species and *P. crassipes* the world distribution is uncertain.

Note: *P. curtipes* and *P. crassipes* are thought to be separate species because of the difference in colour of cell contents mentioned above, difference in sulcal plates and difference in body shape; the apical horn is more prominent in *P. curtipes* than *P. crassipes*.

Protoperidinium decipiens (Jorg.) Parke & Dodge

Fig. 22K

Parke and Dodge in Parke and Dixon, 1976, p. 545.

Syn: *Peridinium decipiens* Jorgensen 1899, p. 40.
                        Lebour, 1925, p. 132, fig. 41d.
                        Schiller, 1937, p. 265, fig. 264.

Cell markedly flattened anterioposteriorly. Epitheca rising abruptly to a low, symmetrical apical horn, with meta-quadra plate arrangement, kidney-shaped in polar view. Sulcus bordered by inconspicuous lists. Girdle right-handed with lists supported by spines. Theca finely reticulated with scattered pores. Nucleus situated on dorsal side, oval.

Size: Length 44–56 µm; breadth 75–90 µm.

Distribution: Rare around British coasts, but reported from the west coast of Norway and Arctic sea.

Protoperidinium divergens (Ehrenb.) Balech

Fig. 22D

Balech, 1974, p. 60.

Syn: *Peridinium divergens* Ehrenberg, 1840, p. 201.
                        Stein, 1883, pl. 10, figs. 1–5, non 6–9.

Lebour, 1925, p. 127, pl. 26, figs. 2a—e
Schiller, 1937, p. 226, fig. 222.

Cell longer than broad with concave sides to both epitheca and hypotheca. Prominent apical horn and two diverging hollow antapical horns. Meta-quadra plate arrangement. Sulcus widening slightly, posteriorly, bordered by lists which project antapically appearing as spines on the inside of the antapical horns. Girdle excavated with lists supported by spines, not offset. Theca strongly reticulated with spines at the junctions of the reticulations. Intercalary striae broad. Cell contents pink. Cells from Millport emitted bright luminescence when stimulated (Tett, 1971).

Size: Length 80—84 μm; breadth 56 μm.

Distribution: Neritic; found in North Sea and off S.W. coast of England, Worldwide distribution unclear because of confusion with *P. curtipes/ P. crassipes.*

Note: There has been much confusion between *P. divergens* and *P. curtipes/crassipes* and *P. depressum. P. divergens* may be distinguished from *curtipes* by its narrower body, although in respect of the plate arrangement and thecal sculpturing they are both similar. It differs from *P. depressum.* by its meta 1' (*P. depressum* has an ortho 1') and its strong sculpturing.

Protoperidinium globulum (Stein) Balech

Fig. 22A

Balech 1974, p. 64.

Syn: *Peridinium globulus* Stein, 1883, pl. 9, figs. 5—7.
        Paulsen, 1908, p. 42.
        Lebour, 1925, p. 129, fig. 40.
        Schiller, 1937, p. 182, fig. 185.

Cell spherical or globular, compressed anterio-posteriorly; apical horn in form of small knob; no other body projections except girdle lists. Meta-quadra or hexa plate arrangement with plates covered in pores. Intercalary striae broad. Girdle strongly right-handed, displaced approximately two girdle widths with slight overhang, not excavated but with lists supported by spines. Sulcus narrow, not reaching to centre of hypotheca but extending onto epitheca. Theca finely reticulated.

Size: 50—79 μm long.

Distribution: Found, rarely, in the English Channel and North Sea. Elsewhere recorded from the Mediterranean, Pacific, Indian and Atlantic Oceans.

Protoperidinium ovatum Pouchet

Fig, 22H–J, Pl. IV f, V b.

Pouchet, 1883, p. 35, pl. 18, fig. 13.
Balech, 1976, p. 35, fig 4

Syn: *Peridinium ovatum* (Pouchet) Schutt, 1895, pl. 16, fig. 49.
                    Lebour, 1925, p. 126, pl. 26, figs. la–d.
                    Balech, 1976, p. 35, figs. 4a–v.
  P. globulus var. *ovatum* Schiller, 1937, p. 186, fig. 187.

Cell lenticular, slightly flattened anterio-posteriorly, hypotheca
ending with two solid spines. Meta-quadra (or sometimes penta) plates on
the epitheca. There is a small apical horn. the girdle is offset; may or may
not be excavated, with lists supported by spines. The hypotheca is rounded
antapically and bears two slightly diverging spines either side of the sulcus
each with two wings. The sulcus is shallow. The thecal plates (and some of
the sulcal plates) are sculptured with poroids or areolae and reticulations.
Cell contents pink. Intercalary striae may be broad. Tett (1971) found
luminescent cells from waters off Millport.

Size: Length 54–68 µm; breadth 57–88 µm.

Distribution: Occurs commonly all around coast of Britain. Also
found in the Mediterranean, Indian Ocean, Atlantic.

Note: There is confusion concerning *P. ovatum, P. quarnerense,
P. globulum* and *P. cerasus,* the relationship to *P. quarnerense* causing the
most difficulty but Balech (1976c) states that these two differ in size and
general shape, the apex, girdle, spines, plate pattern, ornamentation, sulcal
plates and colour of the protoplasm.

Protoperidinium subcurvipes (Lebour) Balech

Fig. 22N, O

Balech, 1974, p. 61.

Syn: *Peridinium subcurvipes* Lebour, 1923, p. 267, fig. 4.
                    Lebour, 1925, p. 133, pl. 27, figs. 3a–e.
                    Balech, 1959, p. 23, pl. 2, figs. 36, 37.

Cell flattened anterioposteriorly, otherwise globular and round in
polar view. Epitheca with short apical horn, meta quadra plate arrangement.
The hypotheca bears a curved spine on the right antapical plate. The sulcus
has a prominent list down the left side and this curves round at the antapex.
The girdle is right-handed, not excavated, with wide lists supported by

spines. A large pusule is present, the nucleus is large and equatorially situated, and the cell contents pinkish.

Size: Length 47 μm; breadth 38–45 μm.

Distribution: Found off N.E. coast of England, from the English Channel, and off Spitzbergen.

Note: This species has been confused with *P. curvipes* which also bears the curved wing-like left sulcal list. They differ however in their epithecal plates, colour and shape. Schiller (1937) thought the two species were conspecific but Balech (1949) considers them distinct species.

Sub-genus: Protoperidinium. Group Metaperidinium, section Pyriformia.

Protoperidinium granii (Ostenf.) Balech

Fig. 23A

Balech, 1974, p. 65.

Syn: *Peridinium granii* Ostenfeld, 1906, p. 15.
                      Paulsen, 1907, p. 15, fig. 18 partim.
                      Lebour, 1925, p. 124, pl. 25, figs. 2a–b.
                      Schiller 1937, p. 189, fig. 188 g–r, non a–f.
                      Balech, 1971, p. 138, pl. 27, figs. 514–519.
          Peridinium sp.   Gran, 1902, p. 188, figs. 13a–c.

Cell squarish with short apical horn and two hollow antapical horns which may each end in a spine. Sides of epitheca straight or slightly convex or concavo-convex. Meta or occasionally para-penta plate arrangement. Hypotheca straight or slightly concave in ventral view. Sulcus narrow and indenting the antapex between the horns with longitudinal lists projecting antapically. Girdle slightly right-handed, excavated with narrow lists supported by spines. Theca finely reticulated or with scattered pores.

Size: Length 49–80 μm; breadth 49–56 μm

Distribution: Neritic. Not commonly found around Britain but has been recorded from the English Channel, North Sea and western coasts. Also found off Iceland, California, southeastern South America and in the Caribbean.

Note: This species has been frequently confused with *P. mite* from which it differs in having hollow antapical horns and a more square body shape in ventral view. The confusion arose when both Paulsen and Pavillard believed they had specimens which they considered similar to *Peridinium* sp. described by Gran (1902, p. 188, fig. 13). Pavillard named his species *Peridinium mite* and Paulsen named his *Peridinium grani.*

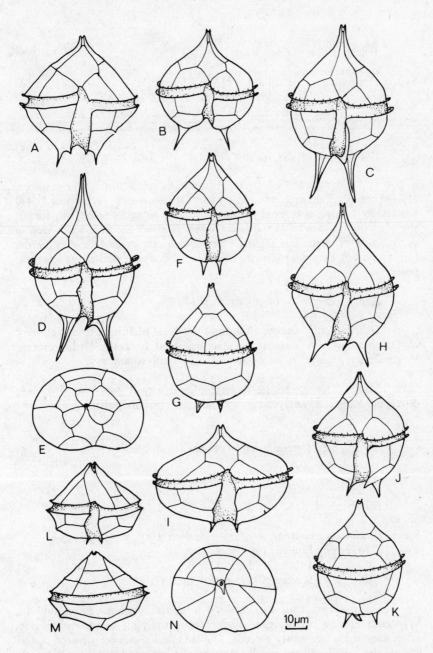

Fig. 23. PROTOPERIDINIUM
A. *P. grani* B. *P. mite* C. *P. steinii* D. *P. diabolum* E. *P. diabolum* apical view F. *P. pyriforme* ventral view G. *P. pyriforme* dorsal view H. *P. pallidum* I. *P. islandicum* J. *P. pellucidum* ventral view K. *P. pellucidum* dorsal view L. *P. thulesense* ventral view M. *P. thulesense* dorsal view N. *P. thulesense* apical view.

Protoperidinium mite (Pav.) Balech

Fig. 23B

Balech, 1974, p. 63.

Syn: *Peridinium mite* Pavillard, 1916, p. 36, figs. 9a–c.
               Lebour, 1925, p. 125, pl. 25, fig. 3.
               Balech, 1971, p. 135, pl. 27, figs. 497–503.
               Balech, 1976, p. 39, figs. 5A–K.

        P. granii f. *mite* Pav. Schiller, 1937, p. 191, figs. 188s–z.

Cell with rounded sides drawn out into an apical horn on the epitheca. Meta-penta plate arrangement; intercalary striae narrow. Hypotheca broad posteriorly bearing two long, slender, diverging antapical spines situated at the edge of the base of the antapex where it meets the sides when seen in ventral view. The girdle is slightly right-handed, not excavated, with narrow lists supported by spines. Theca covered with fine reticulations or scattered pores.

Size: Length 39–60 µm; breadth 34–50 µm.

Distribution: Occurs all around the coast of Britain from February to October but not abundantly. Also recorded from the Mediterranean, Indian Ocean, Atlantic including area off S.E. South America.

Note: This species has been confused with *P. granii* from which it differs in having solid antapical spines and a different body shape.

Protoperidinium pyriforme (Pauls.) Balech

Fig. 23F, G

Balech, 1974, p. 63.

Syn: *Peridinium steinii* var. *pyriformis* Paulsen, 1905, p. 4, figs. 3d–e.
        P. pyriforme Paulsen, 1907, p. 13, fig. 15.
               Lebour, 1925, p. 126, fig. 38.
               Schiller, 1937, p. 194, fig. 191.

Cell pyriform with straight-sided to convex epitheca and a rounded hypotheca bearing two winged spines. Meta-penta plate arrangement; wide intercalary striae frequently present. The antapical spines are situated at the end of the sulcus and are usually straight, the left one being on a line with the sulcus and bearing a wing which continues up the sulcus towards the girdle. Posteriorly the sulcus broadens very slightly. The girdle is excavated shallowly, is bordered by lists supported by spines and is right-handed. Thecal sculpturing formed of fine reticulations.

Size: Length 40–86 µm; breadth 32–50 µm.

Distribution: Oceanic, found all around the coast of Britain. Also from East Greenland, Norway and Australia.

Note: *P. breve (Peridinium steinii* f. *brevis* Paulsen, p. 5, figs. 3a–c, f) was included with *P. pyriforme* by Balech (1973b). They differ in the length of the antapical spines, those of *P. pyriforme* being longer, and the shape of the body, which is broader in *P. breve*, giving it a rhombic appearance.

Protoperidinium steinii (Jorg.) Balech

Fig. 23C

Balech, 1974, p. 63.

Syn: *Peridinium steinii* Jorgensen, 1889, p. 38.
Lebour, 1925, p. 125, pl. 25, figs. 4a–d.
Schiller, 1937, p. 196, fig. 192.
P. michaelis Stein, 1883, pl. 9, figs. 9–14.

Cell pyriform with elongated apical horn and rounded hypotheca bearing two long three-winged spines; round in polar view. Epitheca with meta-penta plate arrangement. Girdle right-handed, not excavated, with prominent lists supported by spines. Chromatophores absent, cell contents colourless or pale pink or yellow. Intercalary striae usually broad. Found by Tett (1971) to be luminescent in samples taken from Millport.

Size: Length including spines 39–60 μm; breadth 22–44 μm.
Length of spines 9–14 μm.

Distribution: Found all around the coast of Britain. Probably oceanic and neritic. Recorded from Atlantic, Pacific and Indian Oceans and Baltic and Caribbean Seas.

Sub-genus: Protoperidinium; Group: Paraperidinium

Protoperidinium curvipes (Ostenf.) Balech

Fig. 22L, M

Balech, 1974, p. 65

Syn: *Peridinium curvipes* Ostenfeld, 1906, p. 15, fig. 128.
Lebour, 1925, p. 135, pl. 29, figs. 1a–c.
Schiller, 1937, p. 201, figs. 197a–j.

Cell roundish-oval with short apical horn and one spine plus a curved wing at antapex. Epitheca conical with a para-quadra plate arrangement which may be variable. The sulcus does not widen markedly towards the antapex and the margins end in a spine. The left spine is larger with a well-developed characteristically curved wing. Girdle slightly right-handed, unexcavated, with lists supported by spines. Cell contents pale yellow. Theca finely reticulated. Plate formula typical for genus.

Size: 44–55 µm long.

Distribution: Found all around coast of Britain, though never in large numbers, also reported from Greenland and Florida.

Note: There is some confusion between this species and *P. sub-curvipes*. Schiller (1937) put them together under *P. curvipes*. Lebour described *P. sub-curvipes* as being flattened anterioposteriorly, with a meta-quadra plate arrangement.

Protoperidinium diabolum (Cleve) Balech

Fig. 23D, E

Balech, 1974, p. 66. 1976d, p. 43, figs. 7a–n.

Syn: *Peridinium diabolus* Cleve, 1900, p. 16, pl. 7, figs. 19–20.
                  Lebour 1925, p. 135, pl. 29, figs. 2a–c.
                  Schiller 1937, p. 204, figs. 198a–h.

Cell pyriform, slightly flattened dorsoventrally with long apical horn and two solid long, diverging antapical spines. Parahexa plate arrangement. The antapical spines are winged, the left one having a small accessory spine at its base directed towards the right. Girdle right-handed (displacement about half girdle width), not excavated but with prominent lists supported by spines. Thecal plates covered with a fine irregular reticulation. The proto-plasm is clear containing a large central to anterio-central nucleus and a large pusule.

Size: Length 50–75 µm; breadth 42–60 µm; spines 20–22 µm.

Distribution: Found sporadically all around coast of Britain. Also found in Atlantic, Mediterranean, Caribbean and off Brazil.

Balech (1976c) discusses the confusion surrounding Cleve's original drawings and subsequent taxonomic difficulties and proposes that his detailed description should serve to fix the characteristics of the species more clearly.

Protoperidinium islandicum (Pauls.) Balech

Fig. 23 I

Balech, 1973, p. 363, pl. 5, figs. 89–100.

Syn: *Peridinium islandicum* Paulsen, 1904, p. 23, fig. 7.
Lebour, 1925, p. 135, fig. 42a.
Schiller, 1937, p. 206, fig. 201.

Cell depressed from above, broad and short. Almost circular in polar view. Epitheca with short apical horn, sides convex near girdle and concave towards apex. Para-hexa plate arrangement, with smooth or poroid theca. Hypotheca with convex sides indented by sulcus at antapex. At antapical margins of sulcus there are spines, one on the right side and one on the left, with a sulcal list projecting to give the appearance of a further spine. The right sulcal list is denticulate. Girdle right-handed, excavated with lists supported by spines. Yellow-brown chromatophores said to be present.

Size: Length 43 µm − 68 µm; breadth 45−83 µm.

Distribution: Rare off the north coast of Britain. Also recorded off Iceland, Greenland and Spitzbergen and by Balech (1973a) from SE South America.

Protoperidinium pallidum (Osten.) Balech

Fig. 23H

Balech, 1973, p. 365, pl. 6, figs. 101−110.

Syn: *Peridinium pallidum* Ostenfeld, 1899, p. 60.
Lebour, 1925, p. 134, pl. 28, figs. 1a−d.
Schiller, 1937, p. 209, fig. 206.
Graham, 1942, p. 32, figs. 42−43.

In ventral view squarish in outline, slightly longer than broad; dorso-ventrally flattened. The epitheca in ventral view is triangular with apex drawn out to give an apical horn. Para-hexa plate arrangement. Hypotheca straight-sided or slightly convex with indentation at sulcus, and long spines situated on the antapex on either side. The right spine is usually longer than the left and both are winged. On the left side there is a transverse list giving the appearance of a further spine. The girdle is slightly right-handed, may or may not be excavated, and is bordered by conspicuous lists supported by spines. Usually pinkish cell contents. When present, intercalary striae may be very broad. Specimens of this species have been reported to be luminescent.

Size: 70−100 µm long; 66−85 µm broad.

Distribution: Occurs frequently all around coast of Britain and can appear at all times of the year. Recorded from Atlantic, Mediterranean, Arctic and Antarctic seas.

Note: This species is very similar in shape and plate pattern to
P. pellucidum Bergh; indeed Gran (1902) combined them. It has been
suggested by Lebour that they differ in size, *P. pellucidum* being smaller, and
in thecal sculpturing, *P. pellucidum* having a reticulate surface and *P. pallidium*
a finely dotted surface. See Graham (1942) for discussion on these points.
If they are later considered to be the same species then the name *P. pellucidum*
would have priority. Around Britain the two types do seem to be distinct
with *P. pellucidum* smaller than *P. pallidum*.

Protoperidinium pellucidum Bergh

Fig. 23J, K, Pl. V c.

Bergh, 1882, p. 227, pl. 15, figs. 46–48.

Syn: *Peridinium pellucidum* Schutt, 1895, pl. 14, fig. 45.
                              Lebour, 1925, p. 134, pl. 28, figs. 2a–d.
                              Schiller, 1937, p. 212, fig. 209.

Cell slightly flattened dorsoventrally with rounded sides in ventral
view. Para-hexa plate arrangement on epitheca. Short apical horn present.
Hypotheca bears two antapical spines which may or may not have lists. When
present they are small. The left side of the sulcus bears an oblique list giving
the appearance of a third spine. The girdle is slightly right-handed, excavated
and bordered by lists supported by spines. Intercalary striae when present
may be broad and the surface finely reticulated. The sulcus widens slightly,
posteriorly. A large pusule is usually seen; chromatophores are absent, but the
cell contents are pinkish, yellowish or colourless.

Size: Length 40–68 µm; breadth 36–70 µm.

Distribution: Neritic and oceanic. Found all around Britain mostly
in Spring and Summer but occurs throughout year. Also found in the Pacific,
Mediterranean and Atlantic oceans and off Greenland.

Note: This species is very similar to *P. pallidum* and they have
been thought, e.g. by Gran (1902), to be conspecific. *P. pellucidum* appears
to be of more delicate appearance with less heavily winged spines.

Protoperidinium thulesense (Balech) Balech

Fig. 23L, N

Balech, 1973b, p. 27.

Syn: *Peridinium thulesense* Balech, 1958, p. 92, pl. 6, figs. 152–160.

P. conicum f. *islandica* Braarud, 1935, p. 108, fig. 27.
P. sympholis Hermosilla & Balech, 1969, p. 9, figs. 1–3.

Conical straight-sided epitheca being longer than the hypotheca which has convex or straight sides and a flattened antapical end. Almost circular in polar view but may be polygonal. This species characteristically has sutures radiating from the apex to the girdle causing difficulty in interpretation of the plate formula. There may be three or four apical plates, the first is triangular in shape with the right-hand part of the apex of the triangle longer than the left and reaching halfway along the pore plate. Plates 6" and 7" are more or less triangular. The sulcus is narrow and straight and the plates have been described by Balech (1958). The girdle is excavated but without lists.

Size: Length 32–40 μm; breadth 34–40 μm.

Distribution: Found off NE coast of Scotland and in the English Channel between April and June. Also recorded off Greenland and the South Sandwich Islands.

Note: The distinct plate pattern of this species, especially the ventral apical plates, is very similar to that of *P. deficiens* (Meunier, 1919) except that 2a is hexa and not quadra. The drawing copied by Schiller (1937) from Woloszynska (1928) as *P. deficiens* has a quadra 2a and a larger epitheca than hypotheca suggesting that this is like Balech's *P. thulesense*. *P. deficiens* as drawn by Meunier has epitheca and hypotheca of equal length.

Family Gonyaulacaceae

AMPHIDOMA Stein

Stein, 1883, p. 9.
Kofoid & Michener, 1911, p. 274.

Body more or less biconical or hypotheca rounded with small hollow projection. Girdle circular or slightly displaced, centrally positioned or slightly towards the posterior causing epitheca to be larger than hypotheca. Sulcus short, not usually reaching antapex, restricted to hypotheca.

Plate formula: 6', Oa, 6", 6"', 1p, 1""

Type species: *A. acuminata* Stein

Amphidoma caudata Halldal

Fig 24A, B

Halldal, 1953, p. 51, fig. 17.
Silva, 1968, p. 38, pl. 5, fig. 2.

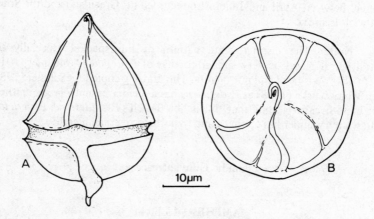

Fig. 24.   AMPHIDOMA

A. *Amphidoma caudata* ventral view B. *Amphidoma caudata* apical view
showing possible arrangement of ridges.

Syn: *Oxytoxum tonollii* Rampi, 1969, p. 51, fig. 1.
        O. margalefi Rampi, 1969, P. 52, fig. 3.

Epitheca characteristically larger than hypotheca and usually twice as high. Epitheca steeply conical, straight-sided or slightly concave in dorso-ventral view. Girdle deeply incised, slightly displaced and not bordered by lists. The hypotheca has gently convex sides and rounded antapex, the latter having a small prolongation on the left side. This prolongation may be short and rounded or elongated, and usually bears a delicate projection. The sulcus broadens posteriorly but does not appear to reach the antapex. The right side of the sulcus may be bordered by a list. The theca is thin and bears no ornamentation and the plate sutures are difficult to discern but ridges do appear to radiate from the apex or from a short distance below it. No plates have been described for the antapex. The very large nucleus is centred usually to the left side of the girdle and extends into both epitheca and hypotheca. Chloroplasts have not been seen.

Size: 24–42 x 19–36 μm.

Distribution: Found off the south and west coast of Britain, east of Scotland and Galway. Previously described from Weather Station M (North Atlantic), the Norwegian coast, from Portuguese waters and the Mediterranean.

Note: This is a small and rather inconspicuous species. It can readily be confused with members of the genera *Katodinium* and *Oxytoxum*.

GONYAULAX Diesing

Diesing, 1866, p. 52.

Variable in shape from spheroidal to polygonal, girdle equatorial or displaced up to seven times its width. Sulcus mainly on hypotheca but may indent epitheca. Theca constructed of delicate to substantial plates: 3–5', 0–2a, 6", 6c, s, 6''', 1p, 1''''; with or without ornamentation. Thecal plates shared after division or shed before cell divides. Resistant cyst stage may occur.

Type species: *G. spinifera* Diesing

Note: This is clearly a composite and therefore very unsatisfactory genus which merges with both *Triadinium (Heteraulacus, Goniodoma)* and *Ceratocorys*. Recent studies have shown that many species have cyst stages the earliest of which to be described was *Spiniferites* (Ehrenberg, 1838). The cysts take a variety of forms and consequently have been placed in a number of distinct cyst genera. As at least three types of cyst, (e.g.: *Spiniferites, Nematosphaeropsis, Tectodinium*) have been linked with motile cells identified as *G. spinifera* it has been suggested that the stage with most

taxonomic significance is the cyst, whereas dinoflagellate taxonomy is traditionally based on the motile stages. Proposals are being put forward for dividing the genus into some eight genera based upon the cysts. However, as these
represent the cysts of only six motile taxa it remains to be seen what will be
done with the 80 or so other described species of *Gonyaulax*. It is also to be
noted that although cyst-genera are valuable for geologists and those studying
sediments they are of very little use in plankton studies for the cysts once
formed rapidly settle to the bottom of the sea. There is clearly a need for
revision, but less drastic than that proposed.

Recent studies have added to our knowledge of the structure of
Gonyaulax thecae and of the life cycle of these organisms in nature. Balech
(1977b) made detailed studies of the thecal plates of four species and suggested
that there is doubt about the homogeneity of the genus. He transferred
G. catenata into the genus *Peridiniella* Kof. & Mich. Taylor (1979) split the
genus more drastically by removing the majority of the toxic species including
G. catenella, G. tamarensis, G. excavata and six others to a new genus
Protogonyaulax. In a paper published in the same volume as that of Taylor,
Loeblich & Loeblich (1979) propose a number of taxonomic changes which
conflict with those described above. For example, *Gonyaulax tamarensis* and
seven other species are transferred to the genus *Gessnerium* Halim.

In view of the considerable confusion which now exists both regarding
the status of the names of the fossil cysts in relation to certain species and the
nomenclatural problems outlined above I have decided to retain the tradit-
ional usage of the genus *Gonyaulax*.

KEY TO THE SPECIES OF GONYAULAX

1a	Thecal plates very delicate with no ridges or reticulations	2
b	Thecal plates thick	3
2a	Cells solitary	*G. tamarensis*
b	Cells forming short chains, may have antapical spines	*G. catenella*
3a	Girdle equatorial, offset by not more than one girdle width	4
b	Girdle ± equatorial, considerably offset	7
4a	Cell with two to four prominent antapical horns/spines	6
b	Cell entire	5

5a	Cell polyhedral in outline; surface of plates poroid	*G. polyedra*
b	Cell ± rounded; plates with pores and strong reticulations	*G. grindleyi*
6a	Cell longer than broad; antapical spines, only two, substantial	*G. verior*
b	Cell width/length ± equal; spines up to four rather small and delicate	*G. triacantha*
7a	Plate ornamentation of reticulations; normally two antapical spines	8
b	Plates covered with longitudinal ridges; no spines or, if present, very feeble	11
8a	Girdle offset only two to two and a half widths; cell longer than broad	9
b	Girdle offset c. three widths; cell about equally as long as broad	10
9a	Antapical horns two; thecal plates prominent and thick	*G. digitale*
b	Antapical horns two to four, small; thecal plates rather delicate	*G. spinifera*
10a	Antapical spines two, rather small; sides of epitheca ± concave	*G. alaskensis*
b	Antapical spines minute; sides of epitheca and hypotheca convex	*G. diegensis*
11a	Cell longer than broad; girdle offset only about one width	*G. polygramma*
b	Cell ± as broad as long; girdle offset two to three widths	*G. scrippsae*

Gonyaulax alaskensis Kofoid

Fig. 25K

Kofoid, 1911, p. 249, pl. 17, figs. 45–46.
Schiller, 1937, p. 304, fig. 315.

Cell rather rhomboidal in shape, length and breadth more or less the same. Epitheca triangular leading to a short apical horn; hypotheca with almost straight sides and a flattened antapex from which project two sulcal-list spines. Girdle offset about three widths, ridged and with narrow lists and large overhang; sulcus sinuous and deeply impressed. Thecal plates randomly reticulate except along either side of the girdle. Said to have chloroplasts.

Size: 65–75 µm in length and width.

Distribution: Found in the North Sea and reported from the English Channel, Mediterranean and Pacific..

Gonyaulax diegensis Kofoid

Fig. 26G

Kofoid, 1911, p. 217, pl. 13, figs. 21–24.
Lebour, 1925, p. 95, pl. 13, fig. 5.
Schiller, 1937, p. 281, fig. 285.

Cell nearly spherical but with an undulating outline, slightly longer than broad. Epitheca with convex sides leading to short apical horn; hypotheca rounded, bearing two to four very short spines. Girdle median, displaced about three widths with only a slight overhang, strongly ridged and excavated but with no lists; sulcus sinuous. Plates thick with reticulations and broad intercalary bands; plate formula 4′, 0a, 6″, c, s, 6‴, 1p, 1″″. Probably contains chloroplasts.

Size: Length 56-100 µm; width 50–82 µm.

Distribution: A sparsely found species around the north and west of the British Isles. Also reported from the Pacific and off Australia.

Cyst: None known.

Gonyaulax digitale (Pouchet) Kofoid

Fig. 26A, Pl. VI c, d.

Kofoid, 1911, p.214, pl. 9, figs. 1–5.
Lebour, 1925, p. 92, fig. 28a.
Schiller, 1937, p. 283, fig. 286.

Syn: *Protoperidinium digitale* Pouchet,   1883, p. 443, pl. 18, fig. 14.

Top-shaped to rhomboidal; a largish sturdy cell. Epitheca straight-sided or slightly concave, tapering to apical horn; hypotheca rounded or

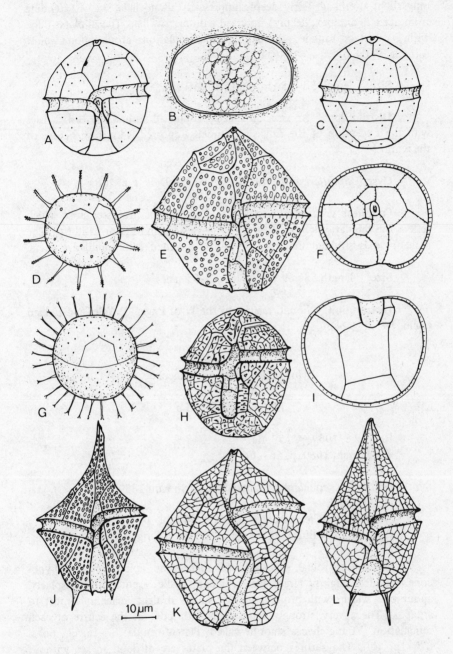

Fig. 25 GONYAULAX
A. *G. tamarensis* ventral view B. *G. tamarensis* resistant cyst C. *G. tamarensis* dorsal view D. *Lingulodinium machaerophorum* cyst E. *G. polyedra* ventral view F. *G. polyedra* anterior view G. *Operculodinium centrocarpum* cyst (and archeopyle) H. *G. grindleyi* ventral view I. *G. grindleyi* antapical view J. *G. triacantha* K. *G. alaskensis* L. *G. verior*.

square-sided with two stout antapical spines. Girdle offset 2-2½ widths with only slight overhang, fairly deeply impressed but no lists; sulcus extending from apex to antapex, deeply impressed with narrow lists. Thecal plates thick with pattern of rounded thickenings or reticulations around pores; plate formula 3', 0a, 6", c, s, 6"', 1p, 1"".

Size: length 50–75 μm; width 35–50 μm.

Distribution: An oceanic species, found all around the British Isles with the exception of the English Channel. Reported from most oceans of the world.

Cyst: *Spiniferites bentori* (Ross.) Wall & Dale. Fig. 26B.

Oval test with two-layered wall from which processes arise which reflect the tabulation (from sutures). Processes end in bifid or trifid tips. The archeopyle is situated in dorsal 3" and is six-sided with rounded corners.

Size: length 58–69 μm; width 52–55 μm.

Distribution: Found mainly to the W of England, Wales and Ireland (Reid, 1972; 1974).

Gonyaulax grindleyi Reinecke

Fig. 25 G–I, Pl. VI b.

Reinecke, 1967, p. 157, fig. 1.
von Stosch, 1969, p. 563, fig. 6–8.

Syn: *Peridinium reticulatum* Claparede & Lachmann 1859, p. 405, pl. 20, fig. 3.
Protoceratium aceros Bergh, 1882, p. 242, pl. 14, fig. 36.
P. reticulatum Butschli, 1885, p. 1007, pl. 52, fig. 2.

Cell subspheroidal, slightly longer than broad; epicone conical, hypo-cone rounded, slightly larger than epicone; girdle ± equatorial, displaced about one width, with prominent lists; sulcus straight, reaching almost to antapex. Theca very strongly reticulated with pore at the centre of each reticulation. Young thecae smooth-walled. Plate formula: 3', 1a, 6", 6c, s, 6"', 1p, 1"". The sutures between the plates are difficult to see without breaking up the theca. Chloroplasts present, brown coloured. Can produce toxin.

Size: length 28–43 μm; width 25–35 μm.

Distribution: All around the British Isles, mainly in summer; said to be neritic but common in the North Sea. Also reported from South Africa.

Cyst: *Operculodinium centrocarpum* (Defl. & Cooks.)
Wall. Fig. 25G.

Spherical and densely ornamented with between 20 and 40 delicate
spines (length is 1/4—1/5 of cell diameter) with capitate and minutely hooked
tips. The yellowish wall is two-layered. The archeopyle is mid-dorsally placed,
precingular, with a trapezoidal aperture.

Size: 33—48 µm diameter.

Described from all around the British Isles, but sparse in the E.
English Channel (Reid, 1974), and from many parts of the world. Recorded
from Quaternary sediments.

Note: This fairly common dinoflagellate was until 1969 known
as *Protoceratium reticulatum,* a name which was abandoned when the taxon
was transferred to the genus *Gonyaulax* as the name *G. reticulata* had already
been used (Kofoid & Mitchener, 1911, p. 271). It has been reported to form
toxic red tides off S.W. Africa (Reinecke, 1967) and can be very abundant in
neritic situations such as Oslofjord (Braarud, 1945).

Gonyaulax polyedra Stein

Fig. 25D—F, Pl. VIa.

Stein, 1883, p. 13, pl. 4, figs. 7—9.
Lebour, 1925, p. 97, pl. 14, fig. 3.
Schiller, 1937, p. 291, fig. 299.

Cell angular or polyhedral in side view with distinct ridges along the
main sutures; almost circular in apical view. Girdle equatorial, displaced
on the right side one to two widths, sulcus running to foot of ventral side.
Thecal plates thick, covered with raised rings at the centre of which are pores;
plate formula 4', 2a, 6", 6c, 8s, 6''', 1p, 1''''. Cell contents brown, chloroplasts
present. Nucleus U-shaped, situated across the centre of the cell. Highly
luminescent.

Size: 42—54 µm diameter.

Distribution: Found in coastal waters along the west of the British
Isles and occasionally in the North Sea. Also reported from temperate and
tropical waters throughout the world.

Cyst: *Lingulodinium machaerophorum* (Deflandre & Cookson) Wall.
Fig. 25D.

Spherical shape with a double-layered wall bearing numerous tapering
spines (up to 9 µm long) which carry tiny spinules near their curved tips. The
archeopyle is large and compound being composed of up to five precingular
plates.

Size: 31–50 μm diameter.

Distribution: Around the British Isles (see Reid, 1972) mainly found in sediments from the E. Irish Sea and the west of Ireland. Present in very small numbers off W. Scotland, SW and SE England.

Note: Many studies have been made of the bioluminescence of this species (e.g. Hastings & Sweeney, 1958). The structure of the cell (Schmitter, 1972) and structure and development of the thecal plates (Durr, 1979) have been studied in detail.

Gonyaulax polygramma Stein

Fig. 26J

Stein, 1883, pl. 4, fig. 15.
Schiller, 1937, p. 292, fig. 300.

Syn: *G. schuetti* Lemmermann, 1899, p. 367.
       Protoperidinium pyrophyrum Pouchet, 1883, p. 433, pl. 18, fig. 15.

Cell elongated with epitheca exceeding hypotheca; girdle descending, displaced about 1.5 widths. Epitheca convex to angular, tapering to the apical horn; hypotheca symmetrical, rounded or truncate, may have a variable number of short antapical spines. Sulcus slightly excavated, widening to the posterior, and to the anterior running onto the epitheca. Plates ornamented with longitudinal ridges which may be very thick and spinulose on old cells; reticulations between the ridges; plate formula 3' (4'), (1a), 6", 6c, 7s, 6"', 1p, 1''''. Chloroplasts present; nucleus large, ovoid, located in the posterior part of the cell.

Size: Length 29–66 μm; width 26–54 μm.

Distribution: Mainly around Scotland, occasionally found in the south west. Also reported from Florida and South Africa.

Cyst: Not known. The pear-shaped cyst with unmarked hyaline pellicle described by Steidinger (1968) would appear to be cytoplasm released prior to division.

Note: There has been confusion here amongst the several members of the 'polygramma' group which includes *G. turbynei* and *G. scrippsae*. This is mainly brought about by the variation in thecal ornamentation which is in part a developmental process (see Steidinger, 1968; Taylor, 1962). Steidinger, after a detailed study, described two extra plates (4', 1a) not reported by earlier workers.

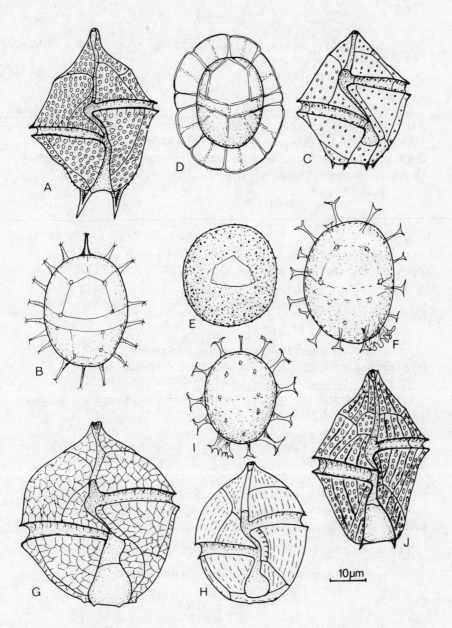

Fig. 26 GONYAULAX
A. *G. digitale* B. *Spiniferites bentori* (cyst) C. *G. spinifera* D. *Nematos-phaeropsis labyrinthula* cyst E. *Tectatodinium pellitum* cyst F. *Spiniferites mirabilis* cyst G. *G. diegensis* H. *G. scrippsae* I. *Spiniferites bulloideus* cyst J. *G. polygramma.*

Gonyaulax scrippsae Kofoid

Fig. 26H, I

Kofoid, 1911, p. 228, pl. 13, figs. 26–27.
Lebour, 1925, p. 94, fig. 28b.
Schiller, 1937, p. 295, fig. 303.

A small rounded organism, with short apical horn. Epitheca with convex sides; hypotheca hemispherical with no spines; girdle median with displacement of 2–3 widths, deeply incised and with strong ridges but no lists. Sulcus angled between ends of girdle, deeply indented. Thecal plates marked by delicate to strong longitudinal ridges or striae; plate formula 3' Oa, 6", c, s, 6"', 1p, 1"".

Size: length 29–50 μm; width 27–40 μm.

Distribution: Found to the SW of the British Isles and also off Yorkshire. Recorded from the Pacific Ocean.

Cyst: *Spiniferites bulloideus* (Deflandre & Cookson) Sarjeant. Fig. 26I.

Oval body ornamented with sutural processes which are fairly long with bifid and trifid tips. Archeopyle situated in the dorsal plate 3".

Size: length 32–42 μm; width 28–39 μm. Length of spines c 16 μm (Wall & Dale, 1968a).

Distribution: Found around most of the British Isles but mainly in the west (Reid, 1974).

Note: Some workers suggest that *G. scrippsae* is a form of the rather variable *G. polygramma* but Wall & Dale, on the basis of the type of cyst, believe that its affinities lie more with *G. spinifera.*

Gonyaulax spinifera (Clap. & Lach.) Diesing

Fig. 26C–F, Pl. 6f.

Diesing, 1866, p. 382.
Lebour, 1925, p. 92, pl. 13, fig. 1.
Schiller, 1937, p. 297, fig. 305.

Syn: *Peridinium spiniferum* Claparede & Lachmann, 1859, p. 405, pl. 20, figs. 4–5.

Cell delicate, top shaped, slightly longer than broad. Epitheca with convex sides leading into an apical horn; hypotheca rounded or with straight sides and two or more small antapical spines. Girdle displaced 2 or more widths with a large overhang, deeply excavated but without lists; sulcus sinuous extending from apex to antapex. Surface of theca variable, may be poroid or delicately reticulate; plate formula 3' 0a, 6", c, s, 6'", 1p, 1''''. Yellow-brown chloroplasts present.

Size: length 24–50 μm; width 30–40 μm.

Distribution: Found all around the British Isles but rare in the English Channel. Reported from most parts of the world.

Cysts: Three distinctive types of cysts have been germinated to give *G. spinifera*-type motile cells (Wall & Dale, 1968a):-

Spiniferites mirabilis (Rossignol) Sarjeant. Fig. 26F.

Oval cell with the antapical area ornamented by processes that are connected by sutural flanges; stout processes ornament the remainder of the test. The wall is thin and two-layered; the archeopyle is dorsal 3", quadrangular with rounded corners.

Size: Width 44–48 μm; length 58–60 μm.

Distribution: Found in small numbers around the British Isles (Reid, 1974).

Nematosphaeropsis labyrinthea (Ostenf.) Reid. Fig. 26D.

Ovoid test surrounded by a series of sutural strands suspended on spines with trifid and bifid tips. Wall of two layers, archeopyle located in 3".

Size: width 30 μm; length 41 μm.

Distribution: Found in small numbers to the N of Ireland and the N and E of Scotland (Reid, 1972).

Tectatodinium pellitum Wall. Fig. 26E.

A spherical cyst without spines. Archeopyle is located in 3". Not reported from around the British Isles.

Note: *G. spinifera* has always been a difficult species to identify with certainty, even with the clear descriptions of Kofoid (1911). There is no sharp dividing line between this species and *G. digitale,* and possibly *G. diegensis* and *G. alaskensis.* There is clearly much variation in the organisms ascribed to this species (cf Schiller, 1937). To make the confusion worse Wall & Dale (1968a) obtained *G. spinifera*-type motile cells from three distinctive types of cyst. This is perhaps the answer to the problem in that *G. spinifera* may be a complex of several species. If it is revised using the resting cysts as the major taxonomic criterion it will mark a major departure in dinoflagellate taxonomy but will also make identification exceedingly difficult for the plankton worker.

Gonyaulax tamarensis Lebour

Fig. 25A–C, Pl. VIe

Lebour, 1925, p. 95, fig. 306.

Syn: *G. excavata* (Braarud) Balech, 1971, p. 28, pl. 7, figs. 119–124.
       G. globosa (Braarud), Balech, 1971, p. 29.
       G. trygvii Parke, 1976, p. 550.
       Pyrodinium phoneus Woloszynskia & Conrad, 1939, p. 3, figs. 1–12.
       Gessnerium tamarensis Loeblich & Loeblich, 1979, p. 44.
       Protogonyaulax tamarensis Taylor, 1979, p. 51.

A small organism with a delicate theca. Cell rounded, slightly longer than broad; girdle equatorial with offset of 0.5–1.5 widths. Epitheca rounded with slight hump at the apical pore; hypotheca rounded but appears slightly depressed where the sulcus, which may be idented, reaches the antapex; sulcal lists may protrude as small spines. Thecal plates smooth with pores; plate formula 4', 0a, 6", 6c, 7–8s, 6"', 1p, 1"". Nucleus U-shaped placed around the centre of the cell; chloroplasts present. May produce toxin and some strains are luminescent.

Size: Length 24–44 μm; width 20–36 μm.

Distribution: A neritic species, around the British Isles has been found mainly in the north and west. Also reported from east U.S.A., where it has formed red tides; Norway, etc.

Cysts: These are subspherical with granular contents, smooth walls, surrounded by mucilage. They form within the theca (Braarud, 1945). The ecological cycle involving cysts has recently been studied (Anderson & Wall, 1978; Turpin *et al.*, 1978).

Note: First described by Lebour (1925), albeit with the epithecal plates reversed, members of what Taylor (1975) has called the 'tamarensis complex' have been found in very large numbers in Europe and the east coast of the U.S.A. Several varieties were described by Braarud (1945) in cultures isolated in Norway. Other workers later raised these to specific rank. Much confusion has been caused by the original erroneous description, the great difficulty of clearly seeing the thecal plates and their variability, and the apparently ephemeral nature of the toxicity. Consequently Taylor (1975b) has suggested that there is perhaps one species which is rather variable. The total variation in the thecal plates which has been reported is 3–5' 0a, 4–6", c, s, 5–7"', 1p, 1-3"". Loeblich & Loeblich (1976) however prefer to retain a number of separate taxa. In the opinion of the present author it is impossible to distinguish clearly between taxa listed above and we therefore follow Taylor (1975b) in this account. See introduction to genus for further discussion of these problems.

Gonyaulax triacantha Jorgensen

Fig. 25J

Jorgensen, 1899, p. 35.

Syn: *Amylax lata* Meunier, 1910, p. 51, pl. 3, figs. 24–27.

A small and delicate species, dorsoventrally flattened; epitheca with concave sides leading into an apical horn with obliquely truncated apex; hypotheca rounded to angular bearing 1–5 or more spines. Thecal plates finely reticulate; plate formula 3', 2a, 6", c, s, 6"', 1p, 1"". Girdle equatorial, with no lists, displaced about one width; sulcus shallow and may extend into an antapical spine. Chloroplasts present.

Size: Length 42–60 μm.

Distribution: Neritic; mainly noted to the north and west of the British Isles and can be common in the North Sea; also reported from Norway, Baltic Sea, Iceland.

Note: This species is easily missed owing to its small size and delicate theca. It can also be confused, particularly with *Peridinium quinquecorne* Abe. It is possible that *G. buxus* (Balech, 1967a) should be included as a synonym, for one-horned forms can be found and have been reported as varieties (Conrad, 1939b).

Gonyaulax verior Sournia

Fig. 25L

Sournia, 1973, p. 34.

Syn: *Amylax diacantha* Meunier, 1919, p. 74, pl. 19, figs. 33–36.
Gonyaulax diacantha (Meun.) Schiller, 1937, p. 300, fig. 309.
G. longispina Lebour, 1925, p. 97, pl. 14, fig. 4.

Non: *Gonyaulax diacantha* Athanassopoulos, 1931.

Cell elongated, inverted top-shaped; epitheca longer than hypotheca, convex sides tapering into an apical horn; hypotheca hemispherical with two conspicuous spines which may be curved with wings, or smaller and almost straight; left spine is longer than the right. Girdle situated below mid-point of cell, offset by 1–2 girdle widths, with narrow lists; sulcus shallow. Thecal plates lightly reticulate; plate formula 4', 2a, 6", c, s, 6"', 1p, 1"". Cell contents yellow, presumably containing chloroplasts.

Size: length 40-56 μm; width c 24 μm.

Distribution: Along the west coast of the British Isles and in the North Sea; previously reported from the Belgian coast and the Mediterranean Sea.

Note: *Amylax diacantha* was first described by Meunier (1919) but Lebour appeared to be unaware of this when she described *Gonyaulax longispina*. The two taxa differed in the length of the antapical spines and the position of the postcingular plates. Examination of material from around the British Isles has shown a range of length and form of the spines. Confusion about the arrangement of the hypothecal plates can readily be caused by slight tilting of the cell. The two species are here brought into synonymy.

Family   Triadiniaceae

TRIADINIUM Dodge

Dodge, 1981

Syn: *Goniodoma* Stein, 1883, p. 12, pl. 7, figs. 1–16.
       Peridinium Pouchet, 1883, partim
       Gonyaulax Okamura, 1907 partim

Non: *Heteraulacus* Diesing, 1850, p. 100.

Cell polygonal or rounded with equatorial girdle which is slightly displaced. Girdle enclosed by conspicuous lists and ridges may be present along the junctions between the plates. Theca composed of regularly punctulate plates with basic formula Po, 3', 7", 6c, 5?s, 5''', 2p, 1''''. There is a distinctive apical pore with a sigmoid aperture. The characteristic feature of this genus is the arrangement of three plates around the apical pore and at the antapex.

Type species: *Triadinium polyedricum* (Pouchet) Dodge

Note: For many years this genus was known under the name *Goniodoma* (Stein, 1883) but this was found to have been used earlier for a genus of insects. Drugg & Loeblich (1967) decided that *Heteraulacus* (Diesing, 1850) was an earlier homonym, however, it is now clear that Diesing's genus which was based on Ehrenberg's *Peridinium acuminatum* referred to an organism quite unlike that described here. Consequently, Dodge (1981) has made the new genus *Triadinium* for these distinctive organisms. Balech (1979), in a recent revision of these organisms, prefers to retain the generic name *Goniodoma*.

Triadinium polyedricum (Pouchet) Dodge

Fig. 27A, B

Dodge, 1981

Syn: *Peridinium polyedricum* Pouchet, 1883, p. 440, pl. 20, fig. 34.
       Goniodoma acuminatum Stein, 1883, pl. 7, figs. 1–16.
       G. polyedricum Jorgensen, 1899, p. 33.
       G. polyedricum Lebour, 1925, p. 90, fig. 26.
       G. polyedricum Schiller, 1937, p. 438, fig. 479.
       Heteraulacus polyedricus Drugg & Loeblich, 1967, p. 183.

Non: *Peridinium acuminatum* Ehrenberg, 1834, p. 541, pl. 2, fig 5.
                              Ehrenberg, 1838, p. 254, pl. 22, fig. 16.
       Heteraulacus acuminatus Diesing, 1850, p. 100.

Cell distinctly polyhedral, seven-sided in ventral view, epitheca with 4 sides, hypotheca with 3; girdle ± circular in apical view. Elongate apical pore directed to the left and surrounded by three almost symmetrical apical plates. Girdle median with wide lists, slightly offset; sulcus deep and bounded by ridges. Plates thick, coverered with pores and with prominent ridges arising from the margins. Plate formula that of the genus. Cell contents dense and chloroplasts are present.

Size: length 40–60 µm; width c 50 µm.

Distribution: Generally regarded as a tropical oceanic species, coming to the British Isles with the Atlantic drift; reported from the Mediterranean sea, Indian Ocean, Caribbean.

Triadinium sphaericum (Murr. & Whitt.) Dodge

Fig. 27C

Dodge, 1981

Syn: *Goniodoma sphaericum* Murray & Whitting, 1899, p. 325, pl. 27, fig. 3.
       G. sphaericum Schiller, 1937, p. 439, fig. 480.
       Heteraulacus sphaericus Loeblich III, 1970, p. 904.

Cell spherical with rounded profile; girdle slightly offset on right, bounded by broad lists; sulcus short. Thecal plates smooth but marked with rows of pores; distinct apical pore present. Plate formula 3', 7", c, s, 5"', 3"". May contain chloroplasts.

Size: 30–50 µm diameter.

Distribution: Found in one sample from the English Channel (December 1970), reported from the Atlantic, China Sea, Caribbean.

Family Heterodiniaceae

HETERODINIUM Kofoid

Kofoid, 1906, p. 342

Syn: *Peridinium* Murray & Whitting, 1899, partim
Gonyaulax Rampi, 1851, partim

Body more or less symmetrical, dorso-ventrally flattened. Apex drawn out into a conical projection or occasionally rounded; antapex forming two antapical spines. Girdle without lists and posterior edge sometimes rather indistinct. Thecal plates thick and strongly sculptured with pores or reticulations. Plate formula 3', 1a, 6", 6c, 6''', 3''''.

This is essentially a genus of tropical waters but occasional specimens are found to the west of the British Isles.

Type species: *H. scrippsii* Kofoid.

Heterodinium milneri (Murr. et Whitt.) Kofoid

Fig. 27F.

Kofoid, 1906, p. 342.

Syn: *Peridinium milneri* Murr. & Whitt. 1899, p. 327, pl. 29, fig. 3.

Cell unequally divided by the girdle, epitheca short and conical, ending in a short apical horn. Hypotheca rounded to squarish with four short antapical spines. Anterior side of the girdle strongly marked, posterior side virtually absent. Thecal plates thick and strongly reticulated.

Size: 50–90 µm long.

Distribution: fairly common in the central Atlantic and occasionally found S.W. of the British Isles (Gaarder, 1954).

Heterodinium murrayi Kofoid

Fig. 27E

Kofoid, 1906, p. 343.

Syn: *Peridinium tripos* Murr. et Whitt. 1899, 327, pl. 30, fig. 4a.

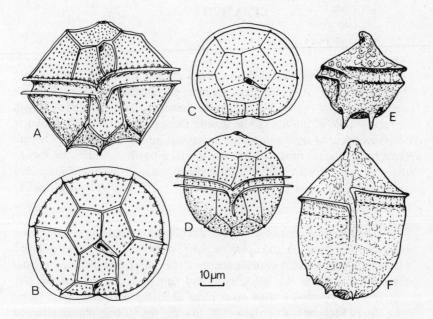

Fig. 27 TRIADINIUM and HETERODINIUM
A. B. *Triadinium polyedricus* C.D. *T. sphaericum* E. *Heterodinium murrayi*
F. *H. milneri*

Similar to *H. milneri* but with longer apical horn and longer antapical spines of which there are here only three of varying lengths. Thecal plates strongly reticulate, girdle displaced.

Size: 35 μm long, 28 μm wide.

Distribution: Tropical Atlantic; one specimen found off Plymouth in 1975. Also found in the Pacific.

Family   Ceratiaceae

CERATIUM Schrank

Schrank 1793, p. 34.

Cell somewhat flattened dorso-ventrally and drawn out into a number (2–4) of hollow horns which may have open or pointed (closed) tips. Epitheca rarely swollen and rounded but generally drawn out into a single apical horn with apical pore at the tip. Girdle slightly descending to the right, ending in a depressed ventral area. Hypotheca drawn out into 2–3 horns, usually with pointed tips. Sulcus modified into an invagination from which the flagella arise. Cell covered with relatively thick thecal plates having numerous trichocyst pores. Basic plate pattern 4', 5'', 4c, 5''', 2'''' plus ventral plates and reduced sulcal plates.

Cells usually contain numerous discoid chloroplasts of yellow-brown colour and may have phagotrophic vacuoles, starch grains and lipid globules. A large nucleus is situated near the centre of the cell. During division daughter nuclei move to lie in the left-anterior and right-posterior ends of the cell, respectively. Cytokinesis then takes place obliquely with the thecal plates being shared between the two new cells. In some species cells remain attached and form chains.

Sexual reproduction is known for several species. The male gametes are normally smaller cells which have often been regarded as forms (e.g. f. *truncata, lata, lineata* in *C. horridum*) (see von Stosch, 1964). The female gametes are of normal size. After fusion the zygote is similar in form to vegetative cells, however, in freshwater ceratia distinctive cysts are formed but no cysts are known from the marine species.

Type species: *C. pleuroceras* Schrank.

The genus is divided into four sub-genera following Sournia (1967):

Sub-genus   Archaeceratium
                   Ceratium
                   Amphiceratium
                   Orthoceratium

Classification of the species of Ceratium
(in the order that they are described)

Sub-genus: ARCHAECERATIUM

C. gravidum

Sub-genus: CERATIUM

C. candelabrum

C. furca

C. lineatum

C. minutum

*C. pentagonum**

*C. teres**

*C. setaceum**

Sub-genus: AMPHICERATIUM

C. fusus

*C. longirostrum**

*C. inflatum**

*C. extensum**

Sub-genus: ORTHOCERATIUM

C. azoricum

*C. petersii**

*C. breve**

C. compressum

C. platycorne

C. tripos

C. arietinum

*C. symmetricum**

*C. gibberum**

C. macroceros

*C. massiliense**

*C. trichoceros**

*C. carriense**

C. hexacanthum

C. longipes

*C. arcticum**

C. horridum

(*brief descriptions only included here)

Note: The form of the cell in this genus is extremely variable and it is often difficult to assign a cell to a definite species. Several statistical studies have been carried out on this problem (e.g. Lopez, 1966; Yarranton, 1967). The variation can be influenced by temperature and season and further confusion comes from the developmental stages following cell division. In the account which follows the numerous forms and varieties which have been described are ignored. The reader is referred to Sournia (1967) and Taylor (1976) for details and references.

KEY TO THE SPECIES OF CERATIUM

1a	Cell consisting of a central body drawn out into two or more slender horns	2
b	Cell body extended into an inflated epitheca or inflated hypothecal horns	5
2a	Hypothecal horns directed in a posterior direction	3
b	Hypothecal horns directed anteriorly or laterally	4
3a	One hypothecal horn long ($\pm$ equal to anterior horn), the other reduced or insignificant (*fusus* group)	6
b	Both hypothecal horns of significant lengths may be equal (*furca* group)	8
4a	Hypothecal horns with pointed tips, epitheca $\pm$ symmetrical (*tripos* group)	20
b	Hypothecal horns with open tips, anterior horn asymmetrically inserted on epitheca (*macroceros* group)	14
5a	Epitheca swollen laterally and flattened, hypotheca extended into two posterior horns	*C. gravidum*
b	Epitheca extended into a horn, hypotheca extended into two open type horns which are swollen	*C. platycorne*

fusus Group (Fusiformia)

6a	Cell over 1 mm (1000 μ) long, hypotheca clearly longer than epitheca, main posterior horn straight	*C. extensum*
b	Cell not normally more than 1 mm long, main posterior horn with slight bend	7
7a	Widest point of cell immediately above girdle followed by rapid constriction, 800–1000 μm long	*C. inflatum*

7b   Widest part of epitheca tapering to about half
length of epitheca, 300–600 µm long   *C. fusus*

c   Epitheca tapering gently from girdle to apex,
500–700 µm long   *C. longirostrum*

furca Group

8a   Main body of cell much broader than long, anterior
horn slightly offset to one side   *C. candelabrum*

b   Main body of cell ± longer than broad, anterior
horn centrally placed   9

9a   Epitheca ± triangular (tapering to apex)
hypothecal horns thick and usually directed
straight backwards   *C. furca*

b   Body of epitheca short and truncated, anterior
horn long with ± parallel sides   10

10a   Anterior horn very long, epitheca about five
times length of hypotheca   *C. setaceum*

b   Epitheca not more than three times length of
hypotheca   11

11a   Posterior horns longer than body of hypocone,
often slightly divergent   *C. lineatum*

b   Posterior horns very short, less than length
of hypocone body   12

12a   Body globular with convex sides to epitheca   *C. minutum*

b   Body pentagonal with straight sides to epitheca   13

13a   Body longer than broad   *C teres*

b   Body wider than or equal to length   *C. pentagonum*

macroceros Group

14a   Body of cell covered with strong reticulations,
antapical horns somewhat twisted   *C. hexacanthum*

b   Body of cell smooth or poroid only, not
markedly twisted   15

15a Hypothecal horns emerging ± laterally, right
    horn arising immediately behind girdle   *C. longipes (arcticum)*

  b Hypothecal horn(s) at least partly directed
    posteriorly   16

16a Right hypothecal horn emerges ± laterally, but becomes
    parallel to anterior horn, both hypothecal horns
    short   *C. horridum*

  b Both hypothecal horns emerge in ± posterior
    direction, long, body of cell small and delicate   17

17a Hypothecal horns almost parallel with anterior
    horn ± equidistant   18

  b Hypothecal horns become lateral/oblique   19

18a Hypothecal horns bend around through 180° to
    become almost as long as anterior horn   *C. macroceros*

  b Hypothecal horns bend sharply through 90° to
    parallel anterior horn, very long and delicate   *C. trichoceros*

19a Hypothecal horns asymmetrical with one much more
    sharply bent round than the other   *C. massiliense*

  b Hypothecal horns widely spread   *C. carriense*

tripos Group

20a Cell very short and squat, flattened dorso-
    ventrally; horns short   21

  b Cell not as above; horns fairly long   22

21a Hypothecal horns become parallel to anterior
    horn   *C. azoricum*

  b Hypothecal horns at slight angle to anterior horn   *C. petersii*

  c Right hypothecal horn considerably longer than
    the left   *C. breve*

22a Cell with straight or irregular posterior margin;
    epitheca with ribbed lists   *C. compressum*

  b Cell with rounded or convex posterior margin   23

23a   Hypothecal horns ± regular curved anteriorly   24

b   Hypothecal horns irregular, apical horn placed
       asymmetrically   25

24a   Hypothecal horns absolutely symmetrical, from
       a convex epitheca, becoming almost as long as
       anterior horn   *C. symmetricum*

b   Right hypothecal horn smoothly curved, left
       rather sharply bent; hypothecal horns not more
       than half length of apical. Body rather coarse   *C. tripos*

25a   Left hypothecal horn may be longer than right;
       anterior horn placed asymmetrically on epitheca   *C. arietinum*

b   Right hypothecal horn twisted, both hypothecals
       short   *C. gibberum*

Sub-genus: Archaeceratium

Ceratium gravidum Gourret

Fig. 28B

Gourret, 1883, p. 58, pl. 1, fig. 15.
Schiller, 1937, p. 357, fig. 388.

Epitheca considerably extended and flattened, oval to round in ventral view, no anterior horn; hypotheca narrowing from the girdle terminating in two straight antapical horns.

Size: Length 300–400 µm.

Distribution: A tropical oceanic species occasionally found in Gulf Stream waters to the west of the British Isles.

Sub-genus: Ceratium

Ceratium candelabrum (Ehrenberg) Stein

Fig. 28A, Pl. VII d

Stein, 1883, pl. 15, figs. 15–16.
Schiller, 1937, p. 364, figs. 401–403.

Syn: *Peridinium candelabrum* Ehrenberg, 1860, p. 792, pl. 1, figs. 2–3.

A distinctive species in which the cell body is broader than it is high. Epitheca obliquely tapering then sharply becoming the straight anterior horn. Hypotheca short and triangular bearing two straight or slightly curved antapical horns, the left being longer than the right.

Size: Breadth 55–70 μm; length (total) 100–200 μm.

Distribution: An oceanic species of warm waters. Found in the Atlantic to the south west of the British Isles during the present study.

Ceratium furca (Ehrenberg) Claparede & Lachmann

Fig. 28C, pl. VIII e

Claparede & Lachmann, 1858, p. 399, pl. 19, fig. 5.
Lebour, 1925, p. 145, pl. 30, fig. 3.
Schiller, 1937, p. 367, figs. 404–5.
Sournia, 1967, p. 395, figs. 18–20.
Taylor, 1976, p. 60, figs. 107–109.

Syn: *Peridinium furca* Ehrenberg, 1836, p. 574, pl. 2, fig. 2.

Body straight, widest either side of the girdle. Epitheca tapering gradually into the anterior horn; hypotheca triangular and slightly tapering, extended into long left and short right antapical horns which are usually straight in line with the cell and may be slightly toothed along the sides. Thecal plates moderately thick, ornamented with a regular reticulum of ridges with pores in the depressions. Central area on the ventral side covered with very delicate plates.

Size: 70–100 μm long; 30–50 μm wide.

Distribution: A cosmopolitan species. Found all around the British Isles.

Ceratium lineatum (Ehrenberg) Cleve

Fig. 28D, Pl. VIII f.

Cleve, 1899, p. 36.
Lebour, 1925, p. 145, fig. 45d, e.
Schiller, 1937, p. 372, fig. 410.
Sournia, 1967, p. 404, figs. 25–26.
Taylor, 1976, p. 61, fig. 118.

Syn: *Peridinium lineatum* Ehrenberg, 1854, pl. 35a, fig. 24c.

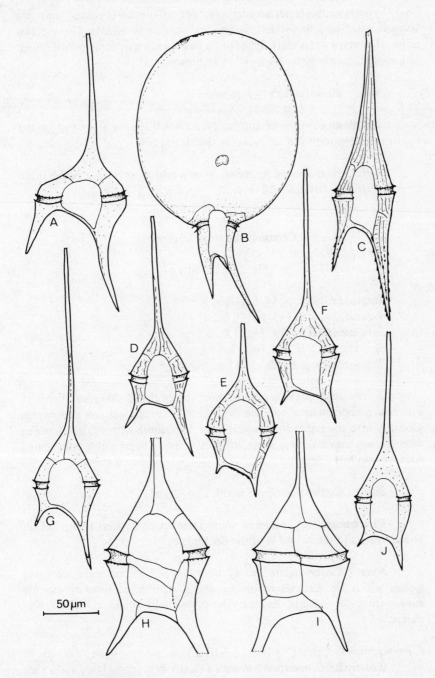

50 μm

Fig. 28 CERATIUM
A. *C. candelabrum* B. *C. gravidum* C. *C. furca* D. *C. lineatum* E. *C. minutum*
F. *C. pentagonum* G. *C. setaceum* H. *C. lineatum* ventral view (after Wall and Evitt 1975) J. *C. teres*

A rather small and delicate member of the genus. Epitheca forming a more or less equilateral triangle with a sharp transition into a fairly long apical horn; hypotheca rather rectangular extended at the lower corners into two unequal antapical horns which are straight but diverge slightly. Thecal plates rather thin and not strongly ornamented but usually have longitudinal ridges. Chloroplasts lightly pigmented, yellowish-brown.

Size: 30–60 μm; 25–45 μm wide.

Distribution: A cool and temperate water species. Found all around the British Isles often very abundant in the N. and N.W.

Note: This species probably merges into *C. minutum* which often cannot clearly be distinguished from it.

Ceratium minutum Jorgensen

Fig. 28E, pl. VIII c

Jorgensen, 1920, p. 34, figs. 21–23.
Lebour, 1925, p. 145, pl. 30, fig. 4.
Schiller, 1937, p. 374, fig. 413.
Sournia, 1967, p. 406, fig. 27.
Taylor, 1976, p. 62, fig. 120.

A very small species but much more sturdy than *C. lineatum*. Epitheca convexly rounded adjacent to the girdle then becoming concave as it sweeps gradually into the rather short apical horn; hypotheca with straight tapering sides leading into two very short antapical horns. Thecal plates ornamented with longitudinal ridges and pores.

Size: length up to 60 μm; width 25–30 μm.

Distribution: A species of warm to temperate waters found in Gulf Stream waters to the N. and W of the British Isles.

Note: Closely related to *C. lineatum* and *C. minutum* are other species which are difficult to distinguish. A number of these species are known from the Atlantic and may be occasionally found in Gulf Stream waters:

C. pentagonum Gourret. (Fig 28F, pl. VIII d.)
   Rather like *C. minutum* but with a clearly pentagonal body and a very long anterior horn.

C. teres Kofoid. (Fig. 28J).
   Similar to the above but with a much more drawn out and triangular epitheca (1½ times the body length).

C. setaceum Jorgensen. (Fig. 28G).

Like the above but with an extremely long anterior spine 3—4 times body length.

Sub-genus: Amphiceratium

Ceratium fusus (Ehrenberg) Dujardin

Fig. 29C

Dujardin, 1841, p. 378.
Lebour, 1925, p. 147, pl. 31, fig. l.
Schiller, 1937, p. 378, fig. 418.
Sournia, 1967, p. 409, figs. 32—34.

Syn: *Peridinium fusus* Ehrenberg, 1834, p. 271.

An elongated needle-shaped species. Widest point adjacent to the girdle from which the epitheca tapers gently into a long apical horn and the hypotheca tapers into a long left antapical horn. A small or rudimentary right antapical horn may be present. Both long horns are usually slightly bent or curved.

Size: 200—300 μm long; 15—30 μm wide.

Distribution: A fairly cosmopolitan species, found all around the British Isles often in large numbers.

Note: Three rather similar species are occasionally found to the W. of the British Isles. They all are normally found in warmer waters:-

C. longirostrum Gourret

Very similar to *C. fusus* but the anterior horn is almost straight and thicker, and the posterior horn perhaps more curved. Epitheca longer than hypotheca. Length c. 250 μm. Found in the E. Atlantic.

C. inflatum (Kofoid) Jorgensen

Similar to *C. fusus* but much larger, length 800—1000 μm. Found in the E. Atlantic.

C. extensum (Gourret) Cleve

Having the same general appearance as *C. fusus* but with the hypotheca markedly longer than the epitheca. The largest species of the series up to 2 mm long. Occasionally found in the E. Atlantic.

Sub-genus: Orthoceratium

Ceratium azoricum Cleve

Fig. 29F

Cleve, 1900, p. 13, pl. 7, figs. 6–7.
Lebour, 1925, p. 151, fig. 48.
Schiller, 1937, p. 406, fig. 447.
Sournia, 1967, p. 435, fig. 58.

Cell short and dorso-ventrally flattened. Anterior horn short and broad; hypotheca rounded with the broad posterior horns pointing in an anterior direction. Posterior horns with pointed tips; wings may be present on apical horn.

Size: length 50–120 μm

Distribution: Said to be a tropical to sub-tropical species. Fairly common in Gulf Stream waters to the W. and N. of the British Isles.

Note: Two species can easily be confused with the above:

C. petersii Nielsen. (Fig. 29E).
This is very similar to *C. azoricum* but is perhaps a little more delicate and may have an angular hypotheca. Size: c. 50 μm in length. Reported from the W. of Ireland.

C. breve (Ostenfeld & Schmidt) Schroder
Again, similar to *C. azoricum* but may be larger (100–200 μm long) and the right posterior horn is usually considerably longer than the left. A species of warm waters.

Ceratium compressum Gran

Fig. 29G, Pl. VIII a.

Gran, 1902, p. 196, fig. 16.
Lebour, 1925, p. 150, fig. 47a.
Schiller, 1937, p. 390, fig. 427b.

Cell rather strongly flattened. Epitheca triangular tapering directly into the short anterior horn which is ornamented with lateral wings. Hypotheca rounded to flattened with the two antapical horns directed anteriorly. These horns are also winged and the right horn commences immediately behind the girdle.

Size: Length 150–200 μm; width of body 50–60 μm.

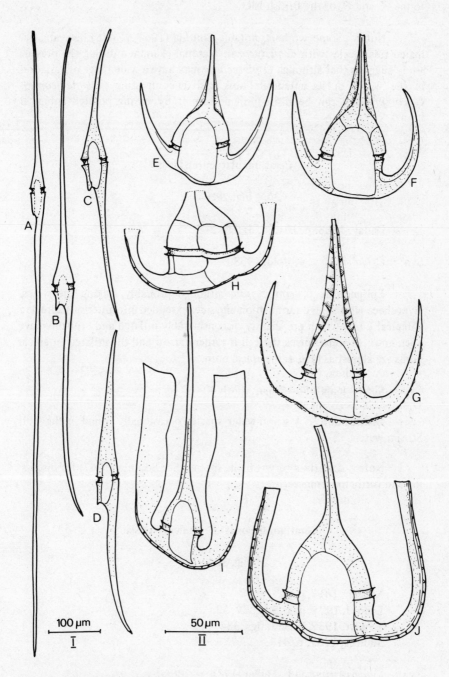

Fig. 29   CERATIUM – Scale I

A. *C. extensum* B. *C. inflatum* C. *C. fusus* D. *C. longirostrum*

Scale II

E. *C. petersii* F. *C. azoricum* G. *C. compressum* H. *C. platycorne* dorsal
view to show plate arrangement I. *C. platycorne* lamellicorne form
J. *C. platycorne*

Distribution: An oceanic species of temperate waters, fairly common to the W. and N. of the British Isles.

Note: Some workers, notably Sournia (1967, p. 455) have amalgamated this species with *C. platycorne*. Material examined during the present study suggests that although *C. platycorne* may have a wide range of variation in the width of its antapical horns (also encompassing *C. lamellicorne)*, *C. compressum* can be distinguished from it by having pointed antapical horns.

Ceratium platycorne Daday

Fig. 29H–J.

Daday, 1888, p. 101, pl. 3, figs. 1–2.

Syn: *C. lamellicorne* Kofoid, 1908.

Epitheca as *C. compressum* although probably lacking the wings. Hypotheca with angled termination and sides extended into anteriorly directed antapical horns which are slightly or considerably inflated and probably have open ends. In some forms the cell is rather narrow and the inflated antapical horns are almost as long as the apical horn.

Size: length 50–70 μm; width 40–60 μm.

Distribution: A warm-water species occasionally found in the Gulf Stream waters.

Note: Clearly a very variable species. See Sournia (1967) for discussion and further references.

Ceratium tripos (O.F. Muller) Nitzsch

Fig. 30A–D

Nitzsch, 1817, p. 4.
Lebour, 1925, p. 148, pl. 32, 33.
Schiller, 1937, p. 382, figs. 384–5.
Sournia, 1967, p. 416.

Syn: *Cercaria tripos* O.F. Muller, 1781, p. 206.

Body of cell stout, about as broad as it is long. Epitheca triangular leading sharply into a fairly long straight horn. Hypotheca slightly flattened at the posterior end, the pointed antapical horns rapidly narrowing and becoming almost parallel with the apical horn. Thecal plates strongly reticulate.

Size: 70–90 μm long

Distribution: A cosmopolitan species, very common around the British Isles.

Note: This species is very variable and numerous varieties have been described (see Sournia, 1967, for references). It probably grades into a number of other species such as those which follow.

Ceratium arietinum Cleve

Fig. 30E, pl. VIII b.

Cleve, 1900, p. 13, pl. 7, fig. 3.
Schiller, 1937, p. 403, fig. 444.
Sournia, 1967, p. 429, figs. 51–54.

Syn: *C. bucephalum* Cleve, 1897, p. 302, fig. 5.
           Lebour, 1925, p. 151, figs. 47b, c.

Epitheca rounded with apical horn placed towards the left side, short and may be winged. Hypotheca very slight, usually rounded at the posterior end, leading directly into rather slender pointed antapical horns which curve round gently or the right may describe more than 180° and become crook-shaped. Thecal plates rather delicately sculptured.

Size: 50–75 μm long; 54–64 μm wide.

Distribution: An oceanic warm-water species, found to the W. and N. of the British Isles in Gulf Stream waters.

Note: *C. symmetricum* Pavillard (Fig. 30G) which has been reported from the E. Atlantic is fairly similar to the above but the antapical horns may converge more strongly and may be longer than the apical horn. *C. gibberum* Gourret (Lebour 1925, p. 152, fig. 49) (Fig. 30F) could also be related to this or *C. tripos.* The anterior horn is asymetrically placed and the antapical horns are much reduced with the right one often bent back to almost touch the epitheca.

Ceratium macroceros (Ehrenberg) Vanhoffen

Fig. 31A

Vanhoffen, 1897, p. 310, pl. 5, fig. 10.
Lebour, 1925, p. 155, (non pl. 35).
Schiller, 1937, p. 428, fig. 468.
Sournia, 1967, p. 460, fig. 83.

Syn: *Peridinium macroceros* Ehrenberg, 1840, p. 201.

A fairly large species (of those found in British Waters) and very delicate. Body small ± pentangular; epitheca short with rounded top leading sharply into the long, slender, apical horn. Hypotheca slightly larger than the epitheca, bearing two long, open-tipped antapical horns which emerge in a posterior direction and then describe an almost perfect half-circle to end more or less parallel with, and almost as long as, the anterior horn. The left antapical horn emerges in line with the apical horn and the right at an angle of about 130°.

Size: Body 50–60 µm long; 45–55 µm wide. Overall length 300–400 µm.

Distribution: Probably a temperate or northern species; found all around the British Isles but more common in the north.

Notes: A number of other species are fairly similar to *C. macroceros* in the small body and very long delicate open horns. Three of these which may occasionally be found in warmer waters to the west of the British Isles are:

C. massiliense (Gourret) Jorgensen. (Fig. 31E).
The main difference here is that the right antapical horn is sharply bent in an anterior direction. Also neither antapical horn turns far in a posterior direction before bending forward. A tropical species reported from the E. Atlantic (Gaarder).

C. trichoceros (Ehrenberg) Kofoid. (Fig. 31 F).
Very similar to *C. macroceros* but the antapical horns emerge almost at right angles to the line of the apical horn and continue for some distance before curving around to lie parallel to the apical horn. A tropical species reported from the W. of Ireland (IMER)

C. carriense Gourret. (Fig. 31 G).
Like a cross between the two preceding species. The antapical horns emerge at right angles to the apical horn and curve in a gentle arc to end at 45° to the apical. Tropical species but has been found W. of Ireland and off Brittany (IMER)

Ceratium hexacanthum Gourret

Fig. 30H, I, Pl. VII e

Gourret, 1883, p. 36, pl. 3, fig. 49.
Schiller, 1937, p 421, fig. 426.
Sournia, 1967, p. 484, fig. 98.

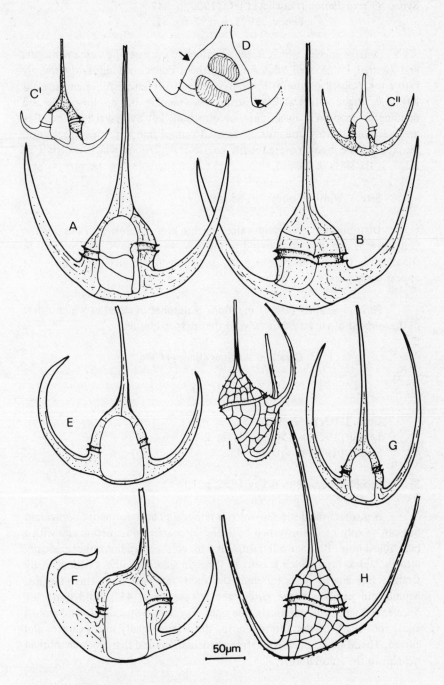

Fig. 30 CERATIUM
A. *C. tripos* ventral view to show the plates (after Wall & Evitt 1975)
B. *C. tripos* dorsal view to show the plates (after Wall & Evitt 1975)
C. I and II recently divided *C. tripos* to show development of new parts
D. *C. tripos* telophase to show plane of cell division E. *C. arietinum*
F. *C. gibberum* G. *C. symmetricum* H. *C. hexacanthum* dorsal view
I. *C. hexacanthum* side view to show twisting

Syn: *C. reticulatum* (Pouchet) Cleve, 1903, p. 342.
                Lebour, 1925, p. 157, fig. 51.

A large species which is quite distinctive because of its ornamentation and twisted horns. Cell body large; epitheca convex leading fairly sharply into a long tapering anterior horn; hypotheca with flat posterior and straight left side. Right antapical horn arising immediately behind the girdle and bending at about 45° in an anterior direction; left antapical horn bending around more sharply and usually twisted out of line with the apical horn. Surface of the body covered with coarse reticulations which protrude as ridges at the sides of the cell.

Size: Width of body 70–85 μm.

Distribution: A warm-water species brought into our area by the Gulf Stream and therefore found in the S.W. and to the N. of Scotland. In the English Channel it can be fairly common in summer but it does not appear every year.

Note: Sournia (1967) mentions a number of varieties which differ in the amount of curvature or twist of the antapical horns.

Ceratium longipes (Bailey) Gran

Fig. 31C, D

Gran, 1902, p. 52, fig. 193.
Lebour, 1925, p. 156, pl. 31, fig. 2.
Schiller, 1937, p. 410, fig. 452.

Syn: *Peridinium longipes* Bailey 1855, p. 12.

A medium-sized species which is rather similar to forms of *C. horridum* but can usually be distinguished by being of coarser construction and with a bent apical horn. Body of cell triangular with epitheca leading rather gradually into the apical horn which is bent to the right side. Hypotheca with virtually straight but angled posterior end; the right antapical horn emerging just behind and parallel to the girdle, the left at about 45°. Both horns are open-ended and bend anteriorly, the right may become parallel to and almost equal the length of the anterior horn, the left is usually much shorter and curved. Thecal plates thick with strong reticulations and there may be toothed margins to the apical horn.

Size: Width of body 50–60 μm; total length up to 250 μm.

Distribution: A species of cooler waters; common around most of the British Isles but appears to be absent S.W. of Ireland.

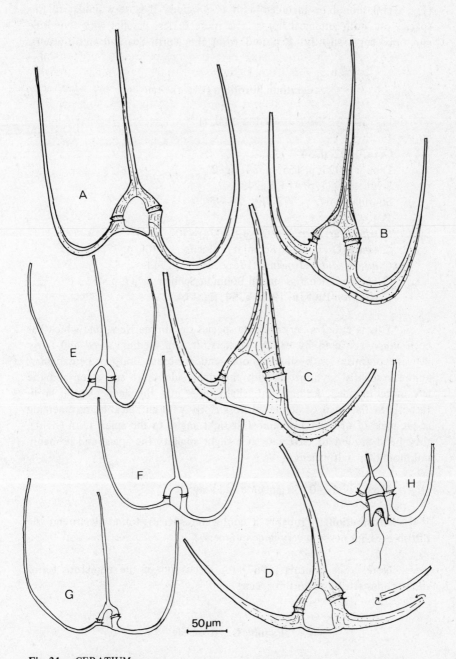

Fig. 31   CERATIUM

A. *C. macroceros* B *C. horridum* C. *C. longipes* D. *C. longipes* f *arcticum*

To half scale

E. *C. massiliense* F. *C. trichoceros* G. *C. carriense* H. *C. horridum* copulation between a macroswarmer and microswarmer (after Von Stosch, 1964)

50μm

Note: Opinions differ as to whether *C. arcticum* (Ehrenberg) Cleve (Fig. 31D) should be included with *C. longipes*. It is very similar to this species but both antapical horns arise more or less parallel with the girdle and only curve slightly. Reported from the North Sea and arctic waters.

Ceratium horridum (Cleve) Gran

Fig. 31B, H

Gran, 1902, p. 54, 193.
Lebour, 1925, p. 155, pl. 34, fig. 2.
Schiller, 1937, p. 413, fig. 455.
Sournia, 1967, p. 477, figs. 91–95.

Syn: *C. tripos* var. *horridum* Cleve, 1897, p. 302, fig. 4.
         C. tenue (Ostenfeld & Schmidt) Jorgensen, 1911, p. 77, fig. 163.
         C. intermedium (Jorgensen) Jorg., 1905, p. 111.
         C. buceros Zacharias emend Bohm in Schiller, 1937, p. 415.
         C. batavum Paulsen, 1908, p. 84, fig. 114.

This is either a very variable species or, alternatively, one which has been much confused by earlier workers. It is a medium sized and fairly delicate organism with shortish open-ended horns. Body ± pentangular; epitheca slightly asymmetric with the right side short and steep and the left more rounded, leading with slight taper into the slender apical horn. Hypotheca flattened or depressed posteriorly with antapical horns emerging at an angle of 45° (left) or almost at right angles to the apical horn (right), they bend around to terminate at a slight angle to the apical and probably not more than half its length.

Size: Body 40–50 μm wide and long.

Distribution: Probably a cool-water species; found all around the British Isles but never in very large numbers.

Note: See Sournia (1967) for discussion of the numerous forms and varieties attributed to this species.

Family Oxytoxaceae

ADENOIDES Balech

Balech, 1956, p. 30.

Cell laterally flattened, with short sulcus on the narrow ventral side of the cell, girdle very close to the anterior end below a much reduced epicone. Cell covered with delicate theca, the epitheca being composed of five plates and hypotheca of 11.

Type species: *A. eludens* (Herdman) Balech

Note: This genus is made up of laterally flattened organisms which were formerly in *Amphidinium*. Following the suggestion of Balech the genus is placed in the family Oxytoxaceae although the justification for this, apart from the much reduced epicone, is not very clear.

Adenoides eludens (Herdman) Balech

Fig. 32A, B

Balech, 1956, p. 30, figs. 1—8.

Syn: *Amphidinium eludens* Herdman, 1922, p. 22, fig. 1.
 A. eludens Lebour, 1925, p. 32, pl. III, fig. 5.
 A. eludens Schiller, 1933, p. 288, fig. 279.

Cell strongly laterally flattened and nearly symmetrical; epicone minute, consisting of a central scarcely discernible protrusion; girdle almost at anterior of cell on ventral side but dipping slightly on dorsal side where it is very shallow; sulcus short enclosed by a flange on its left side. Flagella arise at one third of the length of the cell from the apex, longitudinal flagellum about as long as the body, transverse flagellum encircling anterior end of cell. Cell covered in delicate thecal plates, epitheca of 5, girdle of 5, sulcus of 4, hypotheca of 5''', 5p, 1''''. Cytoplasm containing numerous brown chloroplasts and two pyrenoids. The nucleus is in the dorsal side of the posterior half of the hypotheca, and a large vacuole in the anterior half, opening into the flagellar canal.

Size: 27–30 μm long; 20–25 μm.

Distribution: Originally described from sand at Port Erin, Isle of Man, where it formed brown patches at the north end of the beach (Herdman). Found in the estuary of the River Conway, and on sandy beaches of Norfolk and Brittany.

Adenoides kofoidi (Herdman) comb. nov.

Fig. 32C

Syn: *Amphidinium kofoidi* Herdman, 1922, p. 26, fig. 2.
 A. kofoidi Lebour, 1925, p. 33, fig. 6.
 A. kofoidi Schiller, 1933, p. 299, fig. 293.

Phalacroma Kofoidi Herdman, 1924, p. 79, fig. 23.

Non: *A. kofoidi* var. *petasatum* Herdman, 1924, p. 79, fig. 23.

Body strongly flattened laterally, dorsal side flat and ventral convex; slightly asymmetrical and broadly ellipsoidal. Epicone small button-like, hypocone sac-shaped. Girdle deep, not displaced; sulcus short, bounded by a flange on the left side. Transverse flagellum encircling cell, longitudinal flagellum slightly longer than cell. Cell enclosed in a 'colourless pellicle' (no details of plates have been observed). Cytoplasm containing two pyrenoids, numerous light brown refractile bodies (chloroplasts?) and an elongated nucleus placed in the lower dorsal corner. Anterior part of the hypocone occupied by a clear area (vacuole?).

Size: 25–40 μm long, 25 μm wide.

Distribution: Known only from the sands at Port Erin, Isle of Man.

Note: This organism has not been seen by the present author. From Herdman's description it appears to be remarkably similar to *Adenoides eludens*, particularly as redescribed by Balech (1956), but it is provisionally retained as a separate species pending further attempts to make a new study of it.

CORYTHODINIUM Loeblich & Loeblich III

Loeblich & Loeblich III, 1966, p. 23.
Taylor, 1976, p. 122, fig. 2B.

Syn: *Pyrgidium* Stein, 1883, p. 20.
Murrayella Kofoid, 1907b, p. 191.
Pavillardinium de Toni, 1936, p. 2.
Oxytoxum Stein, partim.

In 1883 Stein assigned five species to his new genus *Pyrgidium* which was similar to *Oxytoxum* but his drawings for these two genera were inverted. The *Pyrgidium* species had five postcingular plates and 1''' was shorter and narrower than the others. The sulcus was well defined and indented and an antapical spine was present. As described by Stein the genus intergraded with *Oxytoxum*, indeed later authors included *Pyrgidium* in *Oxytoxum* (Schutt 1895, Schiller 1937). Kofoid (1907b) created the genus *Murrayella* which was renamed *Pavillardinium* by de Toni since *Murrayella* was preoccupied by the red algal genus *Murrayella* Schmitz. Taylor (1976) in defining the limits of *Oxytoxum, Centrodinium* and *Corythodinium* described the latter as "Relatively robust species, rounded or broadly elongated, lacking "affixed" apical spines (althouth the epitheca may rise to a pointed cone) but usually possessing an antapical spine." The epitheca often shows signs of dextral torsion, with the plates displaced accordingly. In a few species the epitheca may be laterally flattened and raised into a crest. The girdle is strongly developed,

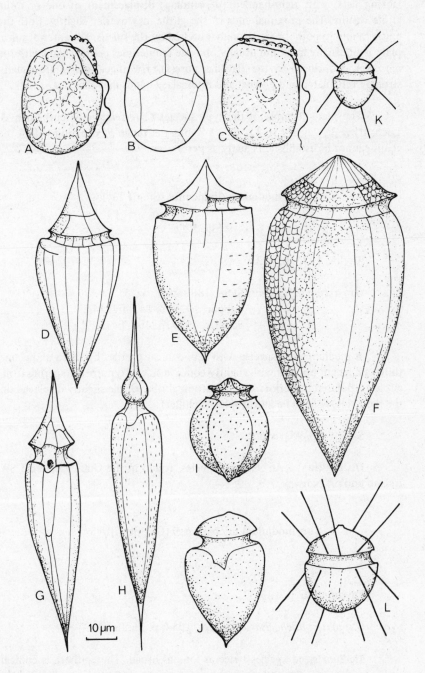

Fig. 32

A. *Adenoides eludens* B. *A. eludens* plates of right side (after Balech 1956)
C. *A. kofoidii* (after Herdman) D. *Corythrodinium diploconus*
E. *C. michaelsarsi* F. *C. reticulatum* G. *Oxytoxum milneri* H. *O. scolopax*
I. *O. sphaeroideum* J. *O. laticeps* K. *Micracanthodinium setifera* (after
Lohman) spines approx. half length L. *M. claytonii* (after Holmes) spines
approx. half length

lacking lists, with right-handed (descending) displacement of one or more girdle widths. The proximal ends of the girdle may overlap slightly. Both the epitheca and hypotheca are strongly notched by the sulcus. The anterior sulcal plate is unusually large and usually obovate (sometimes pentangular) with the narrower end closer to the flagellar pore(s). The thecal plates are usually strongly reticulated and poroid, and intercalary bands may develop.

The plate formula of the type species *C. tesselatum* (Stein) Loeb. & Loeb. III is 3', 2a, 6", 5c ? s, 5"', 1"" but the range of variation within the genus cannot be established clearly as yet.

Corythodinium diploconus (Stein) Taylor

Fig. 32D

Taylor, 1976, p. 123.

Syn: *Oxytoxum diploconus* Stein, 1883, pl 5, fig. 5.
            Lebour, 1925, p. 141, fig. 44d.
            Schiller, 1937, p. 463, fig. 529.

A medium-sized species with a bi-conical profile. Epitheca about one third cell length pointed with slightly concave sides. Hypotheca tapering from girdle to pointed posterior; with longitudinal ribs. Girdle slightly displaced on the right side. Said to be autotrophic (Schiller).

Size: 64–95 µm long.

Distribution: An Atlantic species, found in the Gulf Stream off SW Ireland and off Norway.

Corythodinium michaelsarsii (Gaarder) Taylor

Fig. 32E

Taylor 1976, p. 123.

Syn: *Oxytoxum michaelsarsi* Gaarder, 1954, p. 36, fig. 44.

Medium sized species twice as long as broad. The epitheca is conical with concave sides. Hypotheca with convex sides anteriorly becoming slightly concave towards the antapex which bears a stout horn. The girdle is left-handed, displaced a girdle width, deeply excavated and ribbed. The sulcus is short and narrow. The theca is reticulated and has longitudinal striae crossed by a few faint transverse striae on the hypotheca.

Size: 56 µm long.

Distribution: Found at Weather Station India in the North Atlantic. Also found by Gaarder at several stations of the Michael Sars expedition between Gibraltar and Cape Bojada, and between the Canaries and Sargasso Sea.

Corythodinium reticulatum (Stein) Taylor

Fig. 32F.

Taylor, 1976, p. 123.

Syn: *Pyrigidium reticulatum* Stein, 1883, pl. 5, fig. 14.
　? *Oxytoxum brunellii* Rampi, 1939, p. 466, fig. 24.
　Oxytoxum reticulatum Butschli, 1885, p. 1006, pl. 53, figs. 5–6.
　　　　　　　　Lebour, 1925, p. 142, fig. 44f.
　　　　　　　　Schiller, 1937, p. 462, fig. 525.

Cell elongated with hypotheca at least twice as long as epitheca. Latter short and conical with longitudinal lines radiating from the apex. Hypotheca rounded anteriorly tapering to a point at the antapex. Theca reticulated. Girdle deep and supported by ribs.

Size: 115–120 μm long; Length: breadth = 1 : 0.5.

Distribution: Found at Weather Station India in the North Atlantic. Also from Mediterranean and Atlantic.

OXYTOXUM Stein

Stein, 1883, p. 9.

Cells elongated with pointed apices. Girdle deep and broad, median or anterior to the centre, circular or with slight displacement. Epitheca small, compared with larger epitheca in *Corythodinium*. The anterior sulcal plate slightly indents the epitheca in contrast with *Corythodinium* where it is almost wholly situated in the epitheca. The sulcus itself is short. Thecal plate formula not known since earlier plate formulae were based on larger species which have since been transferred to *Corythodinium*. Plates may have longitudinal sculpturing or poroids. Rarely found in waters around Britain but can occur in the North Atlantic Drift. Typically a warm water genus.

Type species: *O. scolopax* Stein

Oxytoxum laticeps Schiller

Fig. 32J

Schiller, 1937, p. 461, fig. 523.

Cell small, gently rounded at the anterior end, with small papilla;

pointed at the posterior end. Epicone about one quarter the length of the cell like an inverted dish; girdle deeply incised; hypocone helmet-shaped with a small spine. Surface of theca poroid.

Size: length 15—25 μm; width 15—19 μm.

Distribution: Originally described from the Adriatic; found in the E. Atlantic (Station India) and off Norway; also reported from the Indian Ocean and tropical Pacific.

Note: Two other species have a similar shape and size to *O. laticeps:- O. viride* and *O. mucronatum*. There is insufficient information about any of them to be sure of their relationship.

Oxytoxum milneri Murr & Whitt.

Fig. 32G

Murray & Whitting, 1899, p. 328, pl. 27, fig. 6.
Lebour, 1925, p. 141, fig. 44g.
Schiller, 1937, p. 465, fig. 533.
Rampi, 1951, p. 112, fig. 3.
Wood 1968, p. 91, fig. 263.
Steidinger & Williams, 1970, p. 54, fig. 86.

Syn: (?) Oxytoxum challengeroides Kofoid 1907, p. 187, pl. 10, fig. 65.

Cell widest at girdle with epitheca and hypotheca conical, tapering sharply at apex and antapex. Epitheca with long drawn out apical point. Plate margins ridged and plates aereolated. Girdle wide and deep, left-handed, displaced half a girdle width.

Size: 126—131 μm long.

Distribution: Rare in Gulf Stream waters. Found at Weather Station India in N. Atlantic. Found in Atlantic, Mediterranean, Indian Ocean, Tasman Sea, Gulf of Mexico, Caribbean Sea, Straits of Florida and off N. coast of Brazil.

Oxytoxum scolopax Stein

Fig. 32H

Stein, 1883, p. 19. pl. 5, figs. 1—3.
Lebour, 1925, p. 141, fig. 44c.
Schiller, 1937, p. 453, figs. 502a—c.
Balech, 1971a, p. 166, pl. 36, figs. 705—707.
Taylor, 1976, p. 127, pl. 24, figs. 252, 253, pl. 43, fig. 512.

Cell spindle-shaped with acutely pointed ends. Hypotheca about three times the length of the epitheca with an occasional small swelling at the base of the antapical point. Epitheca bulbous at the base and tapering to the apex.

Size: length 70–120 μm

Distribution: A warm-water species which is occasionally found in the Gulf Stream to the south west of the British Isles (Gaarder, 1954). Noted off Ushant and SW of Ireland in the present survey. Also from the Indian Ocean, Gulf of Mexico, Caribbean Sea, Straits of Florida and Sargasso Sea.

Oxytoxum sphaeroideum Stein

Fig. 32 I

Stein, 1883, pl. 5, fig. 9.
Lebour, 1925, p. 140, fig. 44a.
Schiller, 1937, p. 452, fig. 498.

Cell top-shaped, small; epitheca pointed, seated on wide anterior boundary of the wide girdle; hypotheca rounded with pointed antapex. Theca poroid.

Size: length 25–30 μm; width 25 μm.

Distribution: A south Atlantic species but occasionally found to the west of the British Isles.

AMPHIDINIOPSIS Woloszynska 1928

Woloszynska, 1928, p. 256, pl. 7, figs. 1–17.

Syn: *Theacadinium* Balech, 1956, partim.

Cell dorso-ventrally flattened, epicone much smaller than hypocone. Girdle well marked and horizontal except for left-ventral section which runs antapically into the sulcus. Sulcus reaches the antapex and has distinctly curved sides. Thecal plates fairly thick, often with vertical lines of pores or ornamentations. Basic plate formula 3', 7", 5''', 2''''. Type species: *A. kofoidii* Woloszynska.

Note: In the present work two former species of *Thecadinium* are brought into the genus *Amphidiniopsis*. Balech (1956) in describing these species was somewhat uncertain whether they perhaps should be placed in *Amphidiniopsis*. Study of *A. kofoidii* and the type species of *Thecadinium* shows very clearly that they cannot possibly belong to that genus.

Amphidiniopsis kofoidii Woloszynska

Fig. 33B, C

Woloszynska, 1928, p. 256, pl. 7, figs. 1–17.

Cell slightly dorso-ventrally flattened, ellipsoidal in shape. Epicone small, about a quarter cell length, rounded at the apex but with small central notch. Hypocone large, rounded antapically but bilobed in some views due to a sulcal cleft. Girdle distinct, mainly horizontal but the left-ventral portion drops to the centre of the cell where it joins the sulcus. Sulcus either extending the whole length of the cell (Woloszynska) or only from the end of the girdle to the antapex. Curved and broadening at antapex. Thecal plates thick with very fine pores or ornamentation. Thecal plate sutures indistinct but Woloszynska reports 3', Oa, 7", 5'", 2"". The 3'" plate is of a distinctive pentagonal shape and occupies most of the dorsal side of the hypocone, although Woloszynska described it as more rectangular (her figs. 4–6). Cell containing chloroplasts and pigmented granules.

Size: length 30–40 µm; width 23–26 µm.

Distribution: Found in damp sand at Worthing, Sussex; previously reported from the Baltic Sea.

Amphidiniopsis hirsutum (Balech) comb. nov.

Fig. 33A

Syn: *Thecadinium hirsutum* Balech, 1956, p. 40, figs. 38–40.

Cell dorso-ventrally flattened, squarish in shape. Epicone very flattened, triangular, slightly narrower than the hypocone. Hypocone large with a flattened antapex from which numerous small projections arise. Girdle wide and well marked, horizontal except for the angled left-ventral section. Sulcus extending from the girdle to antapex, straight but with curved sides. Thecal plates thick, covered with rows of small projections, sutures wide. Plate formula not clearly established but possibly 4', 3a, 6–7", 5c, s, 5'", 2"". The post-cingular 3'" is rectangular. Cytoplasm colourless.

Size: length 48–51 µm, width 36–44 µm.

Distribution: Found in the English Channel off Dorset. Previously reported from a sandy beach in Brittany.

Amphidiniopsis swedmarkii (Balech) comb. nov.

Fig. 33D, E

Syn: *Thecadinium swedmarki* Balech, 1956, p. 42, figs. 41–42.

Cell strongly dorso-ventrally flattened, a truncated oval in shape. Epicone very small, less than one quarter of cell length, slightly narrower than hypocone, flattened but highest at the central point. Girdle horizontal except for left ventral portion, marked by distinct ridges. Hypocone sac-shaped, large, with two or more short antapical processes. Sulcus running straight from apex

to antapex but with distinctively curved left side. Theca thick, plates 3', 7'',
5''', 2''''. The 3'' is of a distinctive pentagonal shape and centrally placed on
the dorsal side. Sutures sppear to be simple.

Size: length 43–60 μm; width 30–38 μm.

Distribution: Sandy beaches at Studland, Dorset, Port Erin, Isle of
Man and Port Logan, Galloway. First reported from Brittany.

ROSCOFFIA Balech 1956

Balech 1956, p. 44.

Cell more or less ovoid with deeply incised girdle near the anterior end.
Epitheca small and cap-like, hypotheca rounded. Thecal plates present.

Type species: *Roscoffia capitata* Balech

Roscoffia capitata Balech

Fig. 33F

Balech, 1956, p. 44, figs. 43–52.

Small dinoflagellate with slight lateral compression. Epitheca very
small and flattened covered by 4' and 5'' plates, the 1' being extremely narrow.
There is a distinct apical pore. The girdle is wide and deeply incised, slightly
descending on the left side where it runs into a shallow sulcus composed of
three long plates. The pore for the longitudinal flagellum is situated near the
posterior end of the sulcus. The hypotheca is rounded and covered with five
longitudinally orientated plates, two being large lateral plates (2''' and 4''').
All thecal plates are ornamented with reticulations.

Size: 32–34 μm long (epitheca 5 μm), 25–26 μm wide.

Distribution: Originally described from the sandy beaches at Roscoff
(Brittany). As yet, it has not been found around the British Isles.

Note: see Balech (1956) for details of the tabulation.

Family   Cladopyxidaceae

MICRACANTHODINIUM Deflandre

Deflandre, 1937, p. 114.

Small dinoflagellates with more or less median girdle, deeply
excavated, and generally with rounded epi- and hypo-theca. No indication of
thecal plates has been observed. Long, slender unbranched spines project
from the cell and most are inserted in the borders of the girdle.

Type species: *M. setiferum* (Lohmann) Deflandre.

Note: The species in this genus were formerly included in *Cladopyxis* but separated off from that genus on account of the very different thecal structure and unbranched spines. See Balech (1964a) for details of *Cladopyxis* which is not included in the present work.

Micracanthodinium setiferum (Lohmann) Deflandre

Fig. 32K

Deflandre, 1937, p. 114.

Syn: *Cladopyxis setifera* Lohmann 1902, 54, pl. 1 fig. 15

Cell rounded in ventral view, epitheca smaller than hypotheca; girdle wide and fairly deeply excavated. Long fine processes arise from the cell, 4–5 on each half. No details of thecal structure observed. Cell containing golden chloroplasts (Schiller).

Size: 11–12 μm long.

Distribution: Baltic Sea (Kiel), N. Atlantic and the Adriatic.

Micracanthodinium claytonii (Holmes) comb. nov.

Fig. 32L

Syn: *Cladopyxis claytoni* Holmes, 1956, p. 58, figs. 24–25.

Cell with rounded hypotheca and rounded epitheca with a blunt central protuberance. Girdle wide and fairly deeply excavated. Width and length of cell usually equal. Fine spines, 2–3 times cell length, originate from the borders of the girdle, 5–6 on each side. No details of plate structure recorded, the theca dissolves readily in hypochlorite.

Size: length 12–20 μm.

Distribution: Recorded off the coast of Norway and originally described from the Labrador Sea.

Note: The transfer of this species to *Micracanthodinium* was first suggested by Balech (1964a, p. 34).

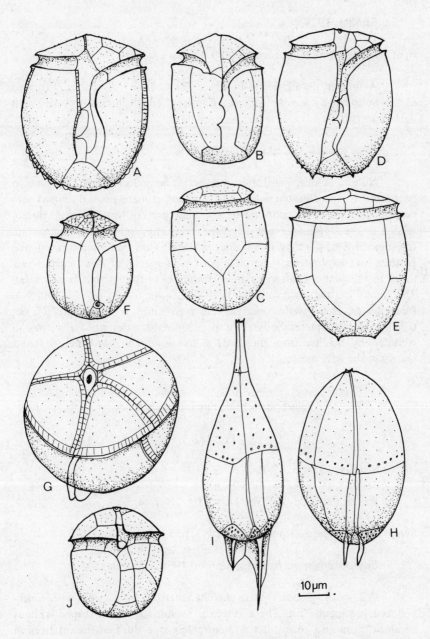

Fig. 33   A. *Amphidiniopsis hirsutum* B. *A. kofoidii* ventral view C. *A. kofoidii* dorsal
view D. *A. swedmarkii* ventral view E. *A. swedmarkii* dorsal view F. *Roscoffia
capitata* (after Balech 1956) G. *Blepharocysta paulsenii* H. *B. splendor-maris*
(after Balech 1963) I. *Podolampas palmipes* (after Balech 1963)
    J. *Palaeophalacroma unicinctum*

PALAEOPHALACROMA Schiller

Schiller, 1928, p. 65.

Syn: *Epiperidinium* Gaarder, 1954, p. 22.

A thecate dinoflagellate in which the girdle is only partially formed and is bounded by a prominent anterior list. Cell is spherical to ovoid with dense contents.

Type species: *P. unicinctum* Schiller.

Note: Schiller considered the species he collected into this genus to be related to the *Dinophysiales*. This was unfortunate since it caused subsequent authors to be confused by the plate pattern which is more closely related to the *Peridiniales* and they then created new genera to accommodate the specimens which they found. Balech (1967b) has clarified the taxonomic position and suggests that the plate pattern of species of *Palaeophalacroma* relate to the genera *Cladopyxis* and *Sinodinium* in the family *Cladopyxidae*. This family has the plate formula 3'–4', Po, 3–4a, 6", 6c, 6''', 2'''' and 5–6 sulcal plates. *Palaeophalacroma* has the plate formula 4', Po, 3a, 7", 6c, 6''', 2'''', 6 sulcal plates. So far, two species have been described, *P. verrucosa* recorded by Schiller from the Adriatic and *P. unicinctum* which is found mainly in the Atlantic.

Palaeophalacroma unicinctum Schiller

Fig. 33J

Schiller, 1928, p. 65, fig. 25.
Schiller, 1933, p. 48, fig. 49a, b.
Balech 1967b, p. 105, fig. 1A–C, fig. 2B, 1–4.

Syn: *Heterodinium detonii* Rampi, 1943, p. 52, figs. 1–6.
　　　　　　　　　　　Halim, 1960, p. 187, pl. 2, fig. 27.
Epiperidinium michaelsarsi Gaarder, 1954. p. 22, fig. 24.

Cell small, globular or ovoid characterized by a girdle with a pronounced precingular list. The epitheca is rounded or dome-shaped without an apical horn and constitutes approximately one third of the total cell in length. The apical pore is slightly ventral in position and is bordered by four apical plates. 1' is long and narrow and posteriorly touches S.a. which extends into the epitheca. There are three intercalary plates and seven precingular plates. These plates may have pores scattered over their surfaces. The cingulum is composed of six plates, is left-handed, displaced by half to one girdle width and slightly excavated. The hypotheca has convex sides and a rounded antapex without any projections. There are six post-cingular plates and two antapicals. The sulcus is not depressed and is formed by six plates. Plate formula: 4', Po, 7", 6c, 6''', 2''''. For a detailed description of these and other thecal plates see Balech (1967b). The nucleus is central, chromatophores are present according to Balech, and the cell contents are rather dense.

Size: 20–42 µm long; 17–37 µm wide.

Distribution: Found at Ocean Weather Station India in the N. Atlantic. Also recorded from the Adriatic and from the S. Atlantic.

Family Podolampaceae

BLEPHAROCYSTA Ehrenberg

Ehrenberg, 1873, p. 4.

Syn: *Peridinium* Ehrenberg, 1859.

Cell spherical to ovoid, girdle apparently absent and sulcus reduced to a narrow groove which may be extended to the apical pore. Cells covered with smooth plates which have a similar tabulation to that of *Podolampas* although there is considerable disagreement over the details of this (see Balech, 1963). The apical and antapical plates are very small and the ranks of larger plates in between probably represent precingulars and cingulars. A genus of warmer waters only occasionally found to the south-west of the British Isles.

Type species: *B. splendormaris* Ehrenberg.

Blepharocysta paulsenii Schiller

Fig. 33G

Schiller, 1937, p. 478, fig. 552.

Cell spherical with slight projection at the apex where there is a prominent apical pore. There are two slight antapical projections from the end of the narrow ventral area. Thecal plates may be thick with wide intercalary bands and are distinctly poroid. Plate formula: 3', la, 5", 4'", 4"".

Size: 28–90 µm diameter.

Distribution: Found in the Atlantic S.W. of the British Isles (Gaarder) and in the Mediterranean.

Note: Gaarder (1954) has provided detailed drawings of the arrangement of the thecal plates.

Blepharocysta splendormaris Ehrenberg

Fig. 33H

Ehrenberg, 1873, p. 4.

Syn: *Peridinium splendor-maris* Ehrenberg 1860, 791.

Cell ovoid or ellipsoidal with two short oar-like processes at the posterior end. Ventral area very narrow (sulcus and 1') leading to apical pore. Thecal plates smooth but poroid, 3', 1a, 5", 3c, 4s, 3'", 3'''' (Balech, 1963).

Size: Length 40–70 μm; width 38–65 μm.

Distribution: In the Atlantic and may be brought to the British Isles by the Gulf Stream. Found off Plymouth May 1975.

PODOLAMPAS Stein

Stein, 1883, p. 9.

This genus contains five or six species which are rarely found around Britain, being typically warm water forms. The genus is characterised by the presence of a distinct apical horn ending in an apical pore, being generally pyriform in shape and bearing two antapical spines. With the genus *Blepharocysta* it shares the unusual feature of the absence of cingular and sulcal depressions.

The plate pattern of *Podolampas* has been interpreted differently by various authors. Kofoid (1909) gave the plate formula as 2', 1a, 6", 3'", 4'''', whereas Balech gave it as 3', 5", 3'", 3''''. For a discussion see Balech (1963).

Type species: *P. bipes* Stein.

Podolampas palmipes Stein

Fig. 33 I

Stein, 1883, pl. 8, figs. 9–11.
Paulsen, 1908, p. 92, fig. 24.
Lebour, 1925, p. 159, fig. 52a.
Schiller, 1937, p. 475, fig. 547.
Balech, 1963, p. 12, pl. 2, figs. 20–27.

Cell pyriform, narrow, ending in two unequal antapical spines. The epitheca is drawn out into a long, slender neck. The intercalary plate is more or less pentagonal. The precingular plates bear elongated pores roughly arranged in a transverse row. There are three cingular plates described by Balech. C_1 and C_3 are long and narrow, C_2 constituting most of the cingular area. The cingulars are notably wide being nearly half the length of the body. The hypotheca is greatly reduced in depth being less than one tenth of body length. The post cingular plates have on them two transverse rows of pores. The first antapical plate (1"") has a large spine which is slightly sinuous and bordered by wings. The plate itself is small and has a few pores. 2"" is small, has pores at the top. 3"" has a short spine bordered by wings. The right-hand spine is therefore much shorter than the left antapical spine. For detailed descriptions of all the plates, including the sulcals, see Balech (1963). The nucleus is elongated and situated in the right half of the cell in the cingulum region. A yellow-orange pigmentation has been recorded by Balech in the epitheca at the base of the neck.

Size: 71—106 μm length of body, 20—37 μm wide. Length of left spine 24—28 μm; length of right spine 14—23 μm.

Distribution: Occasionally found in Gulf Stream waters. Recorded by the present author from Ocean Weather Station India in the North Atlantic. Elsewhere recorded as common in warm waters of the Atlantic and Pacific Oceans.

Order BLASTODINIALES

Family Dissodiniaceae

DISSODINIUM Klebs

Klebs in Pascher, 1916, p. 132.

Syn: *Gymnodinium* Stein, partim.
       Pyrocystis Murray, partim.
       Sporodinium Gonnert.

Dinoflagellates which have a number of distinct stages in the life-cycle with a planktonic cyst-stage most frequently encountered. Motile stages (dinospores) are gymnodinioid, non-armoured, cells. The next stage is parasitic on the eggs of crustacea. Asexual reproduction by production of primary and secondary cysts is followed by formation of dinospores. Primary cyst spherical with central and parietal cytoplasm usually containing chloroplasts. This cyst divides to give 2–16 secondary cysts which are spherical or lunate. Within these 4–16 biflagellate dinospores form. After release these attach to a host wall and insert a haustorial structure. After the contents of the host have been absorbed a primary cyst forms on what remains of the dinospore.

Type species: *Dissodinium pseudolunula* Swift ex Elbrachter & Drebes.

Note: The genera *Dissodinium* and *Pyrocystis* are in a confused state both taxonomically and nomenclaturially. There was initial confusion because both tropical-oceanic (now called *P. lunula* or *D. lunula)* and temperate-neritic (*D. pseudolunula*) organisms form lunate cysts stages. Recent work has shown that these must be distinct species. A second cause of confusion was the lack of knowledge of the life-cycles. Cultural and cinematographic studies of *D. pseudolunula* and *D. pseudocalani* (Drebes, 1974, 1978; Elbrachter & Drebes, 1978) have now resolved this problem. In view of the important parasitic state in the life history of *Dissodinium* this genus has been removed to the Blastodiniales by Elbrachter & Drebes. It is thought that there is no close relationship with the Pyrocystales.

Dissodinium pseudolunula Swift

Fig. 34

Swift ex Elbrachter & Drebes, 1978, p. 362, fig. 1.

Syn: *Gymnodinium lunula* Schutt, 1895, pl. 24, fig. 80, partim.
                    Lebour, 1925, p. 36, pl. 4, fig. 1.
       Pyrocystis lunula sensu Apstein, 1906.
       (for complete listing of the synonymy see Elbrachter & Drebes, 1978)

The life-cycle contains two successive cyst stages leading to dinospore formation. The primary cyst is a spherical ± unpigmented bladder 60–130 µm diameter, in which the protoplasm is peripheral, containing a few plastids, and centrally there is a large vacuole. This cyst divides internally to give 4, 8 or 16 crescent-shaped secondary cysts which are released. These secondary

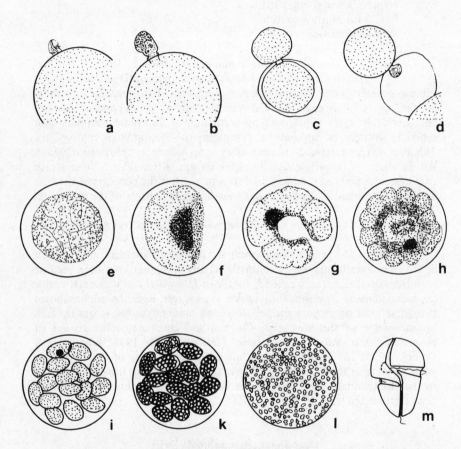

Fig. 34 *Dissodinium pseudolunula* asexual life cycle. *a* young primary cyst filled
with host material; *b* primary cyst, progressive resorption of the host material;
c primary cyst starting nuclear divisions to form secondary cysts; *d*–*e* formation of lunate secondary cysts; *f* release of the secondary cysts by lysis of the
primary cyst wall; *g*–*h* secondary cyst forming dinospores; *i* a dinospore;
k dinospore attached to a copepod egg (host). (Courtesy of Elbrachter &
Drebes 1978)

cysts are the most frequently found stage in the plankton and are 100—140 µm long. The protoplasm retracts from the cell wall at the apices and divides to form a row of 2—16 naked gymnodinioid dinospores which are biflagellate and about 10 x 20 µm in length. These dinospores may contain chloroplasts and their flagella can be seen to be beating before release from the cyst. They lack an armoured theca. Dinospores may infect a host which is usually the eggs of the marine copepod *Temora*. The dinospore attaches by its hypocone and a haustorium is produced which penetrates the egg wall. Feeding takes place and as the contents of the egg are removed so the trophont increases in size. After detachment from the host the trophont develops a wall and becomes the primary cyst stage.

Distribution: common all around the British Isles.

Note: It is quite clear that this is **not** the tropical oceanic organism, usually known as *Pyrocystis lunula*, which is intensely bioluminescent. See p. 141 for possible resting spores of this species.

Dissodinium pseudocalani (Gonnert) Drebes

Fig. 35

Drebes ex Drebes & Elbrachter, 1978, p. 363, fig. 2.

Syn: *Sporodinium pseudocalani* Gonnert, 1936, p. 140, figs. 1—7.

An ectoparasite on eggs of the copepod *Pseudocalanus elongatus.* Colourless swarmers are attached to the copepod eggs where they insert a sucker organ and consume the contents. The parasite then forms a primary cyst which enlarges in size and then divides up to form 8—32 oval secondary cysts by simultaneous nuclear divisions. Subsequently, each secondary cyst divides to give 4—32 colourless biflagellate gymnodinioid swarmers. The organism has none or very few chloroplasts and has a spherical nucleus.

Distribution: North of Scotland; off Helgoland.

Note: See Drebes (1969) and Drebes & Elbrachter (1978) for more complete accounts of the life-cycle.

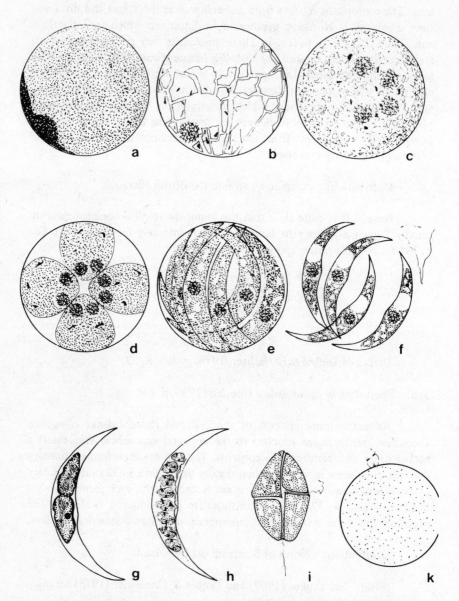

Fig. 35 *Dissodinium pseudocalani* asexual life cycle. *a* dinospore attached to a
copepod egg (host); *b–d* hypertrophic life phase, the dinospore forms a
sucker organelle by which the egg material is incorporated, the parasite swells
to a large spherical bladder; *e* primary cyst; *f–i* formation of secondary
cysts; *k* cyst with 16 secondary cysts just forming dinospores; *l* after lysis
of the secondary cyst walls the dinospores are swarming around inside the
intact primary cyst wall; *m* a dinospore. (Courtesy of Elbrachter & Drebes
1978)

PLATES

LEGEND TO PLATES

Plate I   Prorocentrales

a. *Mesoporus perforatus*, anterior to the right x 3900
b. *M. perforatus* to show a broad megacytic zone between the two main plates x 3100
c. *Prorocentrum micans* x 1830
d. *P. micans*, anterior part to show spine and flagellar pores x 6050
e. *P. balticum*, flagellar pores x 10,450
f. *P. minimum (triangulatum)* x 2600
g. *P. minimum*, anterior end x 8740

Plate II   Dinophysiales

a. *Dinophysis norvegica*   x 1300
b. *D. contracta*   x 1890
c. *D. nasutum*   x 1300
d. *D. caudata*   x 870
e. *D. acuta*   x 580
f. *D. rotundata*   x 580
g. *D. hastata*   x 580

Plate III

a. *Heterocapsa triquetra*   x 3370
b. *Diplopsalis lenticula*   x 1730
c. *D. lenticula*, apical view   x 1300
d. *Scrippsiella trochoidea*   x 2300
e. *S. faeroense*, plate 1'   x 9800
f. *S. faeroense*   x 1660

Plate IV

a. *Protoperidinium cerasus* x 870
b. *P. minutum*   x 1300
c. *P. minutum*, antapical/sulcal view   x 1300
d. *P. brevipes*   x 3110
e. *P. curtipes*   x 730
f. *P. ovatum*, antapical view   x 1300

Plate V

a. *Protoperidinium curtipes*   x 900
b. *P. ovatum*, apical view   x 710
c. *P. pellucidum*   x 870
d. *P. leonis*, dorsal view   x 870
e. *P. depressum*   x 510
f. *P. subinerme*, apical view   x 1340

Plate VI   *Gonyaulax*

a. *Gonyaulax polyedra*   x 1510
b. *G. grindleyi*   x 1300
c. *G. digitale*, dorsal view   x 1300
d. *G. digitale*   x 2600
e. *G. tamarensis*   x 3980
f. *G. spinifera*   x 2590

Plate VII

a. *Triadinium polyedricum* x 700
b. *T. polyedricum*, apical/ dorsal view   x 540
c. *Podalampas bipes*   x 720
d. *Ceratium candelabrum*   x 810
e. *C. hexacanthum*   x 240

Plate VIII

a. *Ceratium compressum*   x 470
b. *C. arietinum*   x 390
c. *C. minutum*   x 1040
d. *C. pentagonum*   x 730
e. *C. furca*   x 860
f. *C. lineatum*   x 580

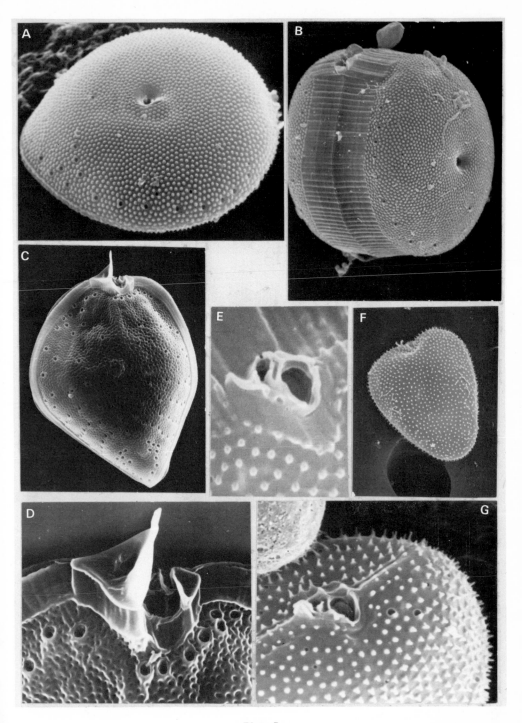

Plate I

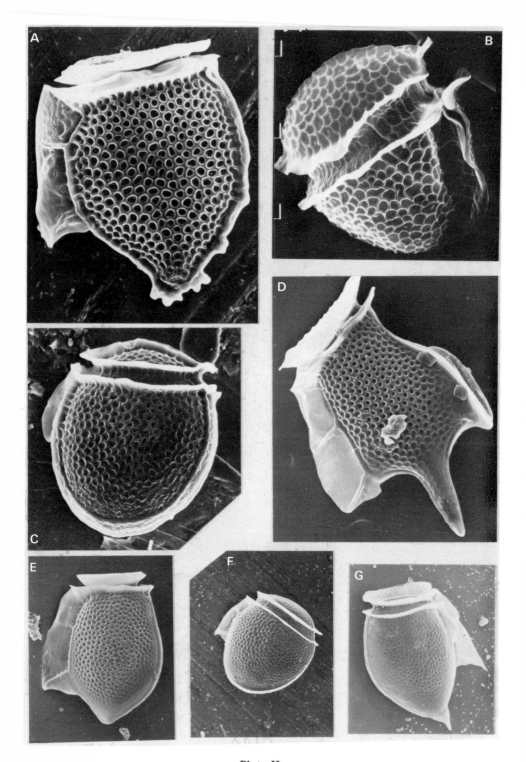

Plate II

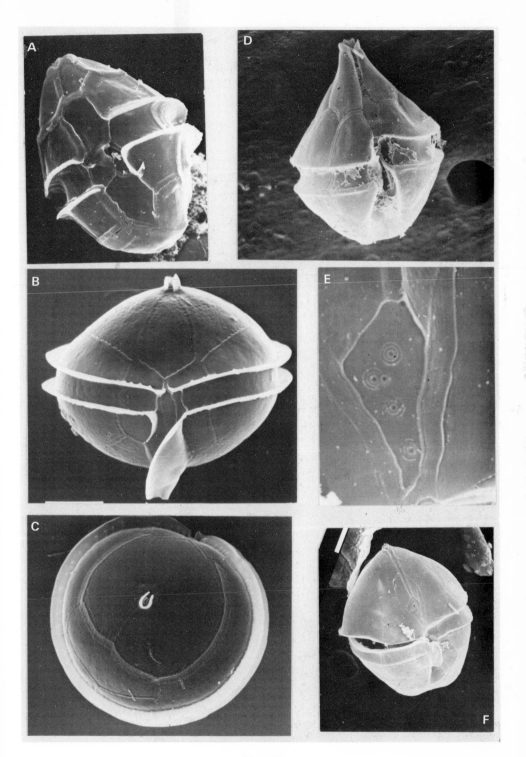

Plate III

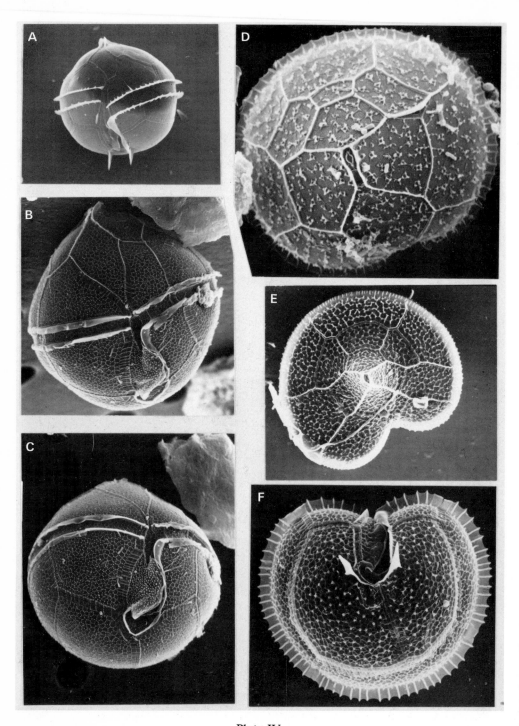

Plate IV

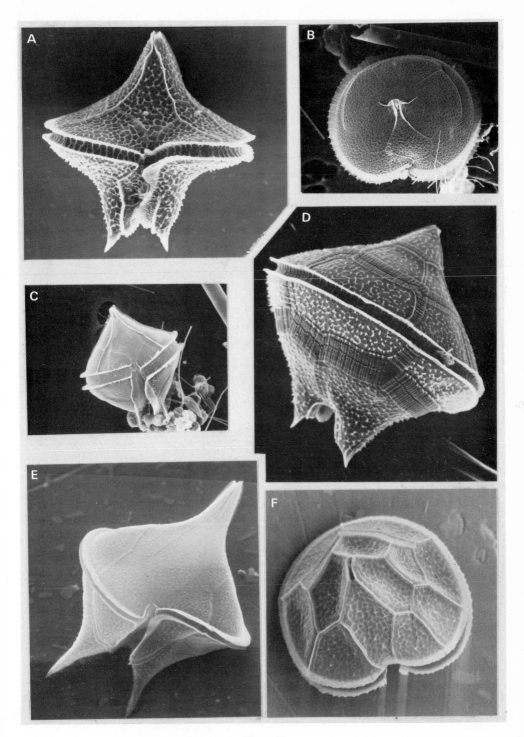

Plate V

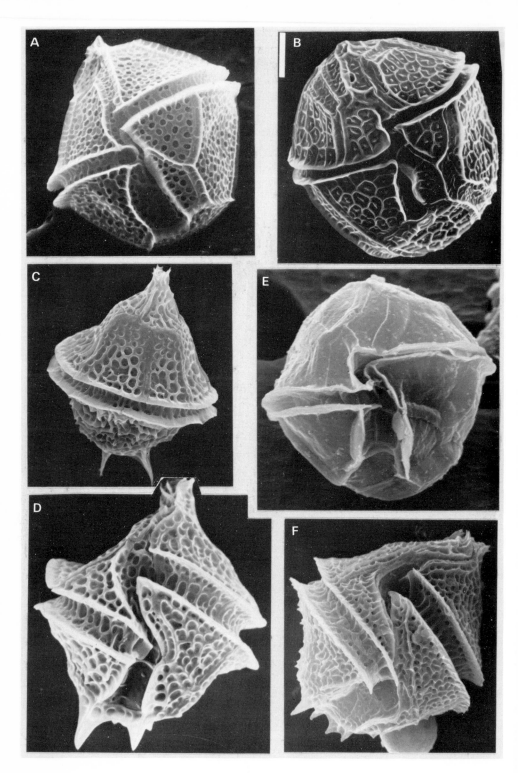

Plate VI

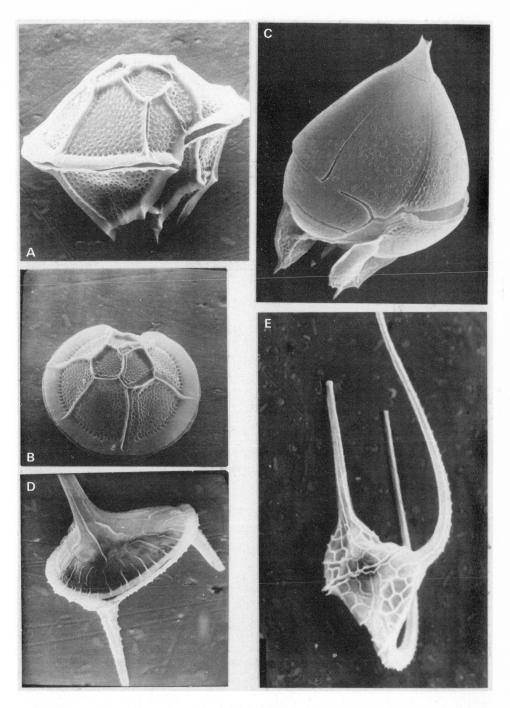

Plate VII

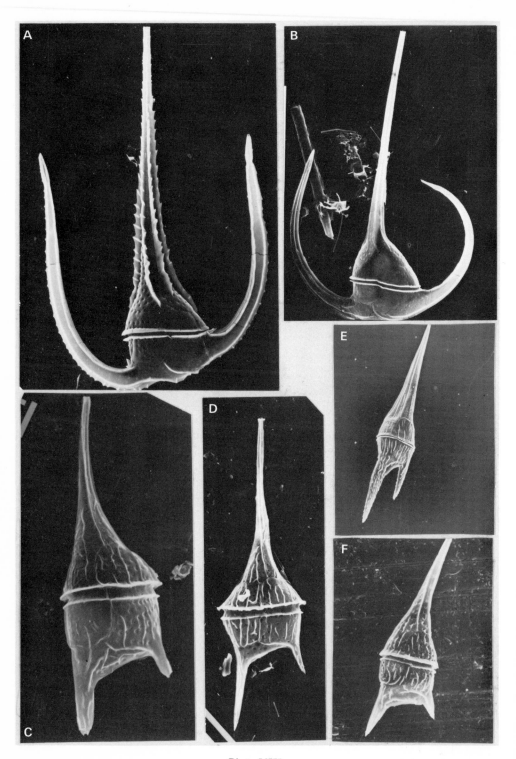

Plate VIII

Acknowledgements

I am grateful to the following persons and organisations who have provided plankton samples or facilities for their collection:—

Mr D. Seaton (DAFS Marine Laboratory, Aberdeen).

Mr P. Wakefield (Tay Estuary Research Centre, Newport on Tay).

Professor M.F. Laverack (The Gatty Laboratory, St. Andrews).

Dr F. Evans (Cullercoats Marine Laboratory, North Shields).

Dr P.C. Head (Department of Civil Engineering, University of Newcastle on Tyne).

Dr J. Lewis (Wellcome Marine Laboratory, Robin Hood's Bay).

Dr S. Coles (Coastal Ecology Research Station, Norwich).

Dr N. Reynolds (MAFF Fisheries Laboratory, Lowestoft).

Mr D. Harding (MAFF Fisheries Laboratory, Lowestoft).

Mrs A. Lincoln (MAFF Fisheries Laboratory, Lowestoft).

Mr J Nichols (MAFF Fisheries Laboratory, Lowestoft).

Mr D. Ayres (MAFF Fisheries Laboratory, Burnham on Crouch).

Mrs M. Cullum (MAFF Fisheries Laboratory, Burnham on Crouch).

Mr P. Davidson (MAFF Fisheries Laboratory, Burnham on Crouch).

Dr W. Farnham (Marine Laboratory, Hayling Island).

Dr P.J.L. Williams (Department of Oceanography, Southampton University).

Dr G. Boalch (Marine Biological Association, Plymouth).

Mr J. Green (Marine Biological Association, Plymouth).

Dr P. Holligan (Marine Biological Association, Plymouth).

Dr P.C. Reid (IMER, Plymouth).

Mr G.A. Robinson (IMER, Plymouth).

Mr N.R. Collins (IMER, Plymouth).

Dr J. Hayward (Botany Department, University College, Swansea).

Dr W.E. Jones (Marine Sciences Laboratory, Menai Bridge).

Dr I. Lucas (Marine Sciences Laboratory, Menai Bridge).

Mrs D. Hepper (MAFF Fisheries Experiment Station, Conway).

Miss F.M. Smith (MAFF Fisheries Experiment Station, Conway).

Mr F. Evans (Lancs. and W. Sea Fisheries Joint Committee, Pollution
and Fisheries Laboratory, Lancaster).

Dr. N. Jones (Marine Laboratory, Port Erin, Isle of Man).

The Director (University Marine Biological Station, Millport).

Dr P. Tett (SMBA, Dunstaffnage Laboratory, Oban).

Dr C. Pybus (Department of Oceanography, Galway, Eire).

Dr I. Jenkinson (Department of Oceanography, Galway, Eire).

REFERENCES

(General literature is listed on p.21)

ABE, T.H. (1927). Report of the biological survey of Mutsu Bay 3. Notes on the protozoan fauna of Mutsu Bay. I. Peridiniales. *Sci. Rep. Tohoku Univ.*, ser *4*, Biol. *2*, 386–438.

ABE, T.H. (1936). II. Genus *Peridinium*: Subgenus *Archaeperidinium*. *Sci. Rep. Tohoku Univ.*, ser. *4*, Biol., *10*, 639–86.

ABE, T.H. (1941). Studies on the protozoan fauna of Shimoda Bay. I. The *Diplopsalis* group. *Rec. oceanogr. Wks. Jap., 12*, 121–44.

ABE, T.H. (1967a) The armoured Dinoflagellata. II. Prorocentridae and Dinophysidae (A). *Publs Seto Mar. Biol. Lab., 14*, 369–89.

ABE, T.H. (1967b). The armoured Dinoflagellata. II. Prorocentridae and Dinophysidae (B). *Dinophysis* and its allied genera. *Publs Seto Mar. Biol. Lab., 15*, 37–58.

ANDERSON, D.M. & MOREL, F.M.M. (1978). Copper sensitivity of *Gonyaulax tamarensis. Limnol. Oceanogr., 23*, 283–95.

ANDERSON, D.M. & WALL, D. (1978). Potential importance of benthic cysts of *Gonyaulax tamarensis* and *G. excavata* in initiating toxic dinoflagellate blooms. *J. Phycol., 14*, 224–34.

BAILLIE, K.D. (1971). A taxonomic and ecological study of the intertidal, sand-dwelling dinoflagellates of the north eastern Pacific Ocean. *M. Sc. Thesis, Univ. Brit. Columbia*, 1–110.

BALECH, E. (1944). Contribucion al conocimiento del Plancton de Lennox y Cabo de Hornos. *Physis, 19*, no. 54, 423–46.

BALECH, E. (1949). Etude de quelques especes de *Peridinium* souvent confondues. *Hydrobiologia, 1*, 390–409.

BALECH, E. (1956), Etude des Dinoflagelles du sable de Roscoff. *Rev. algol.*, (N. ser.) *2*, 29–52.

BALECH, E. (1957). Dinoflagelles et tintinuides de la Terre Adelie (Secteur Francais Antarctiques) recoltes du Dr. Spain-Jaloutre (1950) du Dr. Cendron (1951) et de M. Prevot (1952). Missions Polaires Francaises de P.E. Victor. *Vie et Milieu, 8*, 382–408.

BALECH, E. (1958). Placton de la Campana Antartica Argentina 1954–1955. *Physis 20*, 75–108.

BALECH, E. (1963). La familia Podolampacea (Dinoflagellata). *Bol. Inst. biol. mar., Mar del Plata,* *2*, 1–33.

BALECH, E. (1964a). El genero *"Cladopyxis"* Dinoflagellata. *Revta Mus. argent. Cienc. nat. "B. Rivadavia", Hidrobiol., 1*, 27–39.

BALECH, E. (1964b). El Plancton del Mar del Plata durante el periodo 1961–62 (Buenos Aires, Argentina). *Bol. Inst. biol. mar., Mar del Plata, 4*, 1–49.

BALECH, E. (1967a). Dinoflagelados nuevos o interesantes del Golfo de Mexico y Caribe. *Revta Mus. argent. Cienc. nat. "B. Rivadavia", Hidrobiol., 2*, 77–126.

BALECH, E. (1967b). *"Palaeophalacroma"* Schiller otro miembro de la familia Cladopyxidae (Dinoflagellata). *Neotropica, 13*, 105–112.

BALECH, E. (1971). Microplancton de la campana oceanografica. *Revta Mus. argent. Cienc. nat. "B. Rivadavia", Hidrobiol., 3*, 1–202.

BALECH, E. (1973a). Segunda contribucion al conocimiento del Microplancton del mar de Bellingshausen. *Contr. Inst. antart. argent., 107*, 1–63.

BALECH, E. (1973b). Cuarta contribucion al conocimiento del genero *"Protoperidinium". Revta Mus. argent. Cienc. nat. "B. Rivadavia", Hidrobiol., 3*, 347–68.

BALECH, E. (1974). El genero *"Protoperidinium"* Bergh, 1881 (*"Peridinium"* Ehrenberg, 1831, partim). *Revta Mus. argent. Cienc. nat. "B. Rivadavia", Hidrobiol., 4*, 1–79.

BALECH, E. (1976a). Some Norwegian *Dinophysis* species (Dinoflagellata). *Sarsia, 61*, 75–94.

BALECH, E. (1976b). Notas sobre el genero Dinophysis (Dinoflagellata) *Physis, A 35*, 183–93.

BALECH, E. (1976c). Sur quelques *Protoperidinium* (Dinoflagellata) du Golfe du Lion. *Vie Milieu, 26*, ser. B., 27–46.

BALECH, E. (1976d). Clave illustrada de Dinoflagelados Antarticos. *Publ. Inst. Antart. Argent., 11*, 1–99.

BALECH, E. (1977a). *Cachonina niei* Loeblich (Dinoflagellata) y sus variaciones. *Physis, A 36*, 59–64.

BALECH, E. (1977b). Cuatro especies de *"Gonyaulax"* sensu lato, y consideraciones el genero (Dinoflagellata). *Revta Mus. argent. Cienc. nat. "B. Rivadavia", Hidrobiol., 5*, 115–36.

BALECH, E. (1979). El genero *Goniodoma* Stein (Dinoflagellata). *Lilloa,* *35*, 97–109.

BALLANTINE, D. (1956). Two new marine species of *Gymnodinium* isolated from the Plymouth area. *J. mar. biol. Ass. U.K., 35*, 467–74.

BALLANTINE, D. & SMITH, F.M. (1973). Observations on blooms of the dinoflagellate *Gyrodinium aureolum* Hulburt in the River Conway and its occurrence along the N. Wales coast. *Br. phycol. J., 8*, 233–38.

BARROWS, A.L. (1918). The significance of skeletal variations in the genus *Peridinium. Univ. Calif. Publs. Zool., 18*, 397–478.

BEAM, C.A. & HIMES, M. (1974). Evidence for sexual fusion and recombination in the dinoflagellate *Crypthecodinium (Gyrodinium) cohnii. Nature, 250*, 435–36.

BEAM, C.A. & HIMES, M. (1977). Sexual isolation and genetic diversification among some strains of *Crypthecodinium cohnii*-like dinoflagellates. Evidence of speciation. *J. Protozool., 24*, 532–39.

BERGH, R.S. (1882). Der Organismus der Cilioflagellaten. *Morph. Jb., 7.*, 177–288.

BIECHELER, B. (1952). Recherches sur les Peridiniens. *Suppl. Bull. biol. Fr. et Belg., 36.*, 1–149.

BOALCH, G.T. (1969). The Dinoflagellate genus *Ptychodiscus* Stein. *J. mar. biol. Ass. U.K., 49*, 781–84.

BOLTOVSKOY, A. (1973). *Peridinium gatunense* Nygaard. Estructura y estereoultraestructura tecal (Dinoflagellida). *Physis, B 32*, 331–44.

BOURRELLY, P. (1968). Note sur *Peridiniopsis borgei* Lemm. *Phykos, 7*, 1–2.

BRAARUD, T. (1935). The Phytoplankton and its conditions of growth. The "Ost" Expedition to the Denmark Strait 1929. *Hvalrad. Skrifter, Det norske vidensk. Akad. Oslo, 10*, 1–173.

BRAARUD, T. (1945). Morphological observations on marine dinoflagellate cultures *(Porella perforata, Gonyaulax tamarensis, Protoceratium reticulatum). Vid.-Akad. Avh. I.M. – N.Kgl.* 1944, *11*, 1–18.

BRAARUD, T. & HEIMDAL, B.R. (1970). Brown water on the Norwegian Coast in Autumn. *Nytt Magasin for Botanik, 17*, 91–97.

BRAARUD, T. & PAPPAS, I. (1951). Experimental studies on the Dino-flagellate *Peridinium triquetrum* (Ehrb.) Lebour. *Vid.-Akad. Avh. I.M. – N.Kgl. 1951, 2*, 1–23

BROCH, H. (1910). Die *Peridineen*-Arten des Nordhafens bei Rovigno im Jahre 1909. *Arch. Protistenk., 20*, 176–200.

BURSA, A. (1959). The Genus *Prorocentrum* Ehrenberg. Morphodynamics, protoplasmic structures and taxonomy. *Can. J. Bot., 37*, 1–31.

BURSA, A. (1962). Some morphogenetic factors in taxonomy of Dino-flagellates. *Grana Palynologica, 3*, 54–66.

BURSA, A. (1964). *Kofoidinium arcticum* – a new dinoflagellate. *Phycologia, 4*, 8–14.

CACHON, J. & CACHON, M. (1967). Contribution a l'etude des Noctilucidae Saville-Kent. I. Les Kofoidininae Cachon, J. & M. evolution, morphologique et systematique. *Protistologica, 3*, 427–44.

CARTER, N. (1937). New or interesting algae from brackish water. *Arch. Protistenk., 90*, 1–65.

CHATTON, E. (1933). *Pheopolykrikos beauchampi* nov. gen. nov. sp. Dinoflagelle polydinide autotrophe dans l'etang de Thau. *Bull. Soc. zool. Fr., 58*, 251–54.

CLARKE, K.J. & PENNICK, N.C. (1972). Flagellar scales in *Oxyrrhis marina* Dujardin. *Br. phycol. J., 7*, 357–60.

CLARK, K.J. & PENNICK, N.C. (1976). The occurrence of body scales in *Oxyrrhis marina* Dujardin. *Br. phycol. J., 11*, 345–48.

CONRAD, W. (1926). Recherches sur les Flagellates de nos eaux saumatres. *Arch. Protistenk., 55*, 167–231.

CONRAD, W. (1939a). Notes protistologiques. XIII. *Gonyaulax triacantha* Jorg. var. nov. *subinermis. Bull. Mus. r. Hist. nat. Belg., 15*, (No. 57) 1–3

CONRAD, W. (1939b). Notes protistologiques. X. Sur le schorre de Lilloo. *Bull. Mus. r. Hist. nat. Belg.,15*, (No. 41) 1–18.

CONRAD, W. & KUFFERATH, H. (1954). Recherches sur les Eaux Sumatres des environs de Lilloo. II. Partie descriptive algues et protistes – considerations ecologiques. *Bruxelles, Institut Royal des Sciences Naturelles de Belgique, Memoire No. 127*. pp 1–344

DALE, B. (1977). New observations on *Peridinium faeroense* Paulsen (1905), and classification of small orthoperidinioid dinoflagellates. *Br. phycol. J.*, *12*, 241–53.

DANGEARD, P. (1927). Notes sur la variation dans le genre *Peridinium*. *Bull. Inst. Oceanogr. Monaco*, *24*, 1–16.

DODGE, J.D. (1965). Thecal fine-structure in the dinoflagellate genera *Prorocentrum* and *Exuviaella*. *J. mar. biol. Ass. U.K.*, *45*, 607–14.

DODGE, J.D. (1967). Fine structure of the dinoflagellate, *Aureodinium pigmentosum* gen. et. sp. nov. *Br. phycol. Bull.*, *3*, 327–36.

DODGE, J.D. (1971). A dinoflagellate with both a mesocaryotic and a eucaryotic nucleus. I. Fine structure of the nuclei. *Protoplasma*, *73*, 145–57.

DODGE, J.D. (1972). The ultrastructure of the dinoflagellate pusule: A unique osmo-regulatory organelle. *Protoplasma*, *75*, 285–302.

DODGE, J.D. (1974). A redescription of the dinoflagellate *Gymnodinium simplex* with the aid of electron microscopy. *J. mar. biol. Ass. U.K.*, *54*, 171–77.

DODGE, J.D. (1975a). A survey of chloroplast ultrastructure in the Dino-phyceae. *Phycologia*, *14*, 253–63.

DODGE, J.D. (1977). The early summer bloom of dinoflagellates in the North Sea, with special reference to 1971. *Mar. Biol.*, *40*, 327–36.

DODGE, J.D. (1981). Three new generic names in the Dinophyceae: *Herdmania, Sclerodinium,* and *Triadinium* to replace *Heteraulacus* and *Goniodoma. Br. phycol. J.*, *16*, (3).

DODGE, J.D. & BIBBY, B.T. (1973). The Prorocentrales (Dinophyceae). I. A comparative account of fine structure in the genera *Prorocentrum* and *Exuviaella. Bot. J. Linn. Soc.*, *67*, 175–87.

DODGE, J.D. & CRAWFORD, R.M. (1968). Fine structure of the dino-flagellate *Amphidinium carteri* Hulburt. *Protistologica*, *4*, 231–42.

DODGE, J.D. & CRAWFORD, R.M. (1969). Observations on the fine structure of the eyespot and associated organelles in the dinoflagellate *Glenodinium foliaceum. J. Cell Sci.*, *5*, 479–93.

DODGE, J.D. &CRAWFORD, R.M. (1970). A survey of thecal fine structure in the Dinophyceae. *Bot. J. Linn. Soc.*, *63*, 53–67.

DODGE, J.D. & CRAWFORD, R.M. (1971). Fine structure of the dino-flagellate *Oxyrrhis marina.* I. The general structure of the cell. *Protistologica*, *7*, 295–303.

DODGE, J.D. & CRAWFORD, R.M. (1972). Fine Structure of the dinoflagellate *Oxyrrhis marina*.
    II. The Flagellar System. *Protistologica, 7*, 399–409.

DODGE, J.D. & CRAWFORD, R.M. (1974). Fine structure of the
dinoflagellate *Oxyrrhis marina*.
    III. Phagotrophy. *Protistologica, 10*, 239–44.

DODGE, J.D. & HART-JONES, B. (1974). The vertical and seasonal distri-
    bution of dinoflagellates in the North Sea. *Bot. Marina, 17*, 133–17.

DODGE, J.D. & HART-JONES, B. (1977). The vertical and seasonal distri-
    bution of dinoflagellates in the North Sea. II. Blyth 1973-74 and
    Whitby 1975. *Bot. Marina, 20*, 307–11.

DRAGESCO, J. (1965). Etude cytologique de quelques flagelles mesop-
    sammiques. *Cah. Biol. mar., 6*, 83–115.

DREBES, G. (1974). *Marines Phytoplankton*. Thieme, Stuttgart.

DREBES, G. (1978). *Dissodinium pseudolunula* (Dinophyta) a parasite on
    copepod eggs. *Br. phycol. J., 13*, 319–27.

DREBES, G. & ELBRACHTER, M. (1976). A checklist of Planktonic
    diatoms and dinoflagellates from Helgoland and List (Sylt), German
    Bight. *Botanica mar., 19*, 75–83.

DRUGG, W.S. & LOEBLICH, A.R. (1967). Some Eocene and Oligocene
    phytoplankton from Gulf Coast, U.S.A. *Tulane Studies in Geology,
    5*, 181–94.

DURR, G. (1979). Elektronenmikroskopische Untersuchungen am Panzer
    von Dinoflagellaten. I. *Gonyaulax polyedra. Arch. Protistenk., 122*,
    55–87.

EATON, G.L. (1980). Nomenclature and homology in peridinialean dino-
    flagellate plate patterns. *Palaeontology, 23*, 667–688.

EHRENBERG, C.G. (1834). Dritter Beitrage zur Erkenntniss grosser
    Organisationen in der Richtung des Kleinsten Raumes. *Abh. dt.
    Akad. Wiss. Berl.*, 145–336.

EHRENBERG, C.G. (1838). *Die Infusionsthierchen als volkommene
    Organismen*. Leipzig.

ELBRACHTER, M. (1975). Taxonomical notes on North Sea Dino-
    flagellates. *Kieler Meeresforsch., 1*, 58–64.

ELBRACHTER, M. & DREBES, G. (1978). Life cycles, phylogeny and
    taxonomy of *Dissodinium* and *Pyrocystis* (Dinophyta). *Helgoland
    wiss, Meers., 31*, 347–66.

FOTT, B. (1967). Taxonomic drobnohledni flory nasich vod (Incl. *Katodinium*). *Preslia, 29*, 278–319.

FREUDENTHAL, M.D. (1962). *Symbiodinium* gen. nov. and *Symbiodinium microadriaticum* sp. nov., a Zooxanthella: Taxonomy, Life Cycle and Morphology. *J. Protozool., 9*, 45–52.

FUKUYO, Y., KITTAKA, J. & HIRANO, R. (1977). Studies on the cysts of marine dinoflagellates. I. *Protoperidinium minutum* (Kofoid) Loeblich. *Bull. plankt. Soc. Japan, 24*, 11–18.

FUNG, Y.C. & TROTT, L.B. (1973). The occurrence of a *Noctiluca scintillans* (Macartney) induced red tide in Hong Kong. *Limnol. Oceanogr., 18*, 472–76.

GAARDER, K.R. (1954). Dinoflagellatae from the "Michael Sars" North Atlantic deep-sea expedition 1910. *Rep. scient. Res. 'Michael Sars' N. Atl. Deep-Sea Exped. 1910, 2*, 1–62.

GOCHT, von H. & NETZEL, H. (1974). Rasten elektronenmikroskopische Untersuchungen am Panzer von *Peridinium* (Dinoflagellata). *Arch. Protistenk., 116*, 381–410.

GOUGH, L.H. (1908). Report on the plankton of the English Channel in 1904 and 1905. *Int. Invest. Mar. Biol. Ass. Report, 2*, 165–267.

GRAHAM, H.W. (1942). Studies in the morphology, taxonomy and ecology of the Peridiniales. *Carnegie Inst. of Washington, 542*, 1–127.

GRAN, H.H. (1902). Das Plankton des Norwegischen Nordmeeres von biologischen und hydrographischen Gesichtpunkten behandelt. *Rep. Norweg. Fish. Invest., 2*, pt. 2, No. 5.

GREUET, C. (1971). Etude ultrastructurale et evolution des cnidocystes de *Nematodinium,* peridinien *Warnowiidae* Lindemann. *Protistologica, 7*, 345–55.

GREUET, C. (1972). Le criteres de determination chez les Peridiniens *Warnowiidae* Lindemann. *Protistologica, 8*, 461–69.

GREUET, C. (1978). Organisation ultrastructurale de l'ocelloide de *Nematodinium.* Aspect phytogenetique du photorecepteur des Peridiniens *Warnowiidae* Lindemann. *Cytobiologie, 17*, 114–136.

GRONTVED, J. (1952). Investigations on the phytoplankton in the southern North Sea in May 1947. *Meddr Kommn Danm. Fish. og Havunders., Ser. Plankton, 5*, 1–49.

HASTINGS, J.W. & SWEENEY, B.M. (1958). A persistent diurnal rhythm of luminescence in *Gonyaulax polyedra. Biol. Bull., 115*, 440–58.

HERDMAN, E.C. (1922). Notes on dinoflagellates and other organisms causing discolouration of the sand at Port Erin II. *Trans. Lpool Biol. Soc., 36*, 15–30.

HERDMAN, E.C. (1923). Notes on dinoflagellates and other organisms causing discolouration of the sand at Port Erin III. *Trans. Lpool Biol. Soc.,37*, 58–63.

HERDMAN, E.C. (1924). Notes on dinoflagellates and other organisms causing discolouration of the sand at Port Erin IV. *Trans. Lpool Biol. Soc., 38*, 75–84.

HERMOSILLA, J.G. (1973). Contribution al conicimiento sistematico de los dinoflagelados y tintinidos del archipielago de Juan Fernandez. *Bol. Soc. Biol. Concepcion, 46*, 11–36.

HOLLIGAN, P.M. & HARBOUR, D.S. (1977). The vertical distribution and succession of phytoplankton in the western English Channel in 1975 and 1976. *J. mar. biol. Ass. U.K., 57*, 1075–93.

HULBURT, E.M. (1957). The Taxonomy of the Unarmoured Dinoflagellates of shallow embayments on Cape Cod Massachusetts. *Biol. Bull., 112*, 196–219.

JAVORNICKY, P. (1962). Two scarcely known genera of the class Dinophyceae: *Bernardinium* Chodat and *Crypthecodinium* Biecheler. *Preslia, 34*, 98–113.

JEFFREY, S.W., SIELICKI, M. & HAXO, F.T. (1975). Chloroplast pigments in dinoflagellates. *J. Phycol., 11*, 374–84.

JØRGENSEN, E. (1911). Peridiniales: *Ceratium. Bull. Trim. Bur. Cons. Perm. Int. Expl. Mer. Copenhagen. 7*, 205–248.

JØRGENSEN, E. (1912). Bericht uber die von der schwedischen hydrographisch biologischen Kommission in den schwedischen Gewassern in den Jahren 1909–10 eingesammelten Planktonproben. *Svenska hydrogr.-biol. Komm. Skr., 4*, 1–20.

JØRGENSEN, E. (1923). Mediterranean Dinophysiaceae. *Rep. Dan. Oceanogr. Exped. Mediter., 2*, 1–48.

KAT, M. (1977). Four years phytoplankton investigations in the Dutch Coastal area 1973–76. *Int. Com. Explor-Sea Plank. Cttee. L.2.*

KIMBALL Jr. J.F.& WOOD, E.J. (1965). A dinoflagellate with characters of *Gymnodinium* and *Gyrodinium. J. Protozool., 12*, 577–80.

KOFOID, C.A. (1907a). Dinoflagellata of the San Diego region. III. Descriptions of new species. *Univ. Calif. Publ. Zool. 3*, 299–340.

KOFOID, C.A. (1907b). New species of dinoflagellates. *Bull. Mus. comp. Zool. Harv., 50*, 163–207.

KOFOID, C.A. (1909). The morphology of the skeleton of *Podolampas. Arch. Protistenk., 70*, 48–61.

KOFOID, C.A. & SKOGSBERG, T. (1928). The Dinoflagellata : the Dinophysoidae. *Mem. Mus. comp. Zool. Harv., 51*, 1–766.

KOFOID, C.A. & SWEZY, O. (1921). The free-living unarmoured Dinoflagellata. *Mem. Univ. Calif., 5.* pp. 1–564

KUBAI, D.F. & RIS, H. (1969). Division in the dinoflagellate *Gyrodinium cohnii* (Schiller). A new type of nuclear reproduction. *J. Cell Biol., 40*, 508–28.

LEADBEATER, B. & DODGE, J.D. (1966). The fine structure of *Woloszynskia micra* sp. nov., a new marine dinoflagellate. *Br. phycol. Bull., 3*, 1–17.

LEADBEATER, B. & DODGE, J.D. (1976). An electron microscope study of dinoflagellate flagella. *J. gen. Microbiol., 46*, 305–14.

LEBOUR, M.V. (1917). Microplankton of Plymouth Sound from the region beyond the breakwater. *J. mar. biol. Ass. U.K., 11*, 133–82.

LEBOUR, M.V. (1922). Plymouth peridinians. I. *Diplopsalis lenticula* and its relatives. *J. mar. biol. Ass. U.K., 12*, 795–812.

LEBOUR, M.V. (1925). *The Dinoflagellates of Northern Seas.* Marine Biological Association, Plymouth.

LEE, R.E. (1977). Saprophytic and phagocytic isolates of the colourless heterotrophic dinoflagellate *Gyrodinium lebourae* Herdman. *J. mar. biol. Ass. U.K., 57*, 303–15.

LEMMERMANN, E. (1904). Das Plankton schwedischen Gewasser. *Ark. Bot., 2*, 1–209.

LINDEMANN, E. (1918). Mitteilungen uber Posener Peridineen. *Ebenda, 25*, pt. 1.

LOEBLICH, A.R. III (1965). Dinoflagellate nomenclature. *Taxon, 14*, 15–18.

LOEBLICH, A.R. III (1970). The amphiesma or dinoflagellate cell covering. *Proc. N. Amer. Paleont. Conv. Chicago, 1969.* G, 867–929.

LOEBICH, A.R. III & LOEBLICH, L.A. (1979). The systematics of *Gonyaulax* with special reference to the toxic species. In: *Toxic Dinoflagellate Blooms.* Ed. D.L. Taylor & H.H. Seliger. Vol. I of Developments in Marine Biology. pp. 41–56.

LOEBLICH, A.R. & LOEBLICH, A.R. III. (1970). Index to the genera, sub-genera, and sections of the Pyrrhophyta, IV. *J. Paleont., 44*, 536–543.

LOEBLICH, L.A. & LOEBLICH, A.R. III (1975). The organism causing New England red tides: *Gonyaulax excavata. Proc. of the First Int. Conf. on Toxic Dinoflagellate Blooms, Nov. 1974. Boston, Mass.* Ed. V.R. LoCicero, Massachusetts Science & Technology Foundation, Wakefield, Mass. pp. 207–224

LOEBLICH, A.R. III & SHERLEY, J.L. (1979). Observations on the theca of the motile phase of free-living and symbiotic isolates of *Zooxanthella microadriatica* (Freudenthal) comb. nov. *J. mar. biol. Ass. U.K., 59*, 195–205.

LOPEZ, J. (1966). Variacion y regulacion de la forma en el genero *Ceratium. Invest. Pesq., 30*, 325–427.

MANGIN, L. (1911). Sur l'existence d'individus dextres et senestres chez certains Peridiniens. *C. r. hebd. Seanc. Acad. Sci., Paris, 153*, 27–32.

MATZENAUER, L. (1933). Die Dinoflagellaten des Indisches Ozeans. *Bot. Arch., 35*, 437–510.

MEUNIER, A. (1910). Microplancton des Mers de Barents et de Kara. *Duc d'Orleans, Campagne Arctique de 1907.*

MEUNIER, A. (1919). Microplancton de la mer Flamande. 3. Les Peridiniens. *Mem. Mus. r. Hist. nat. Bruxelles, 8*, Nr. 1.

MEUNIER, V. & SWIFT, E. (1977). Observations on the thecate swarmers of *Pyrocystis* species in laboratory culture: *Pyrocystis fusiformis* Wyville Thomson ex Murray and *Pyrocystis pseudonoctiluca* Wyville Thomson ex Murray (Dinococcales). *Phycologia, 16*, 359–65.

MOREY-GAINES, G. & RUSE, R.H. (1980). Encystment and reproduction of the predatory dinoflagellate, *Polykrikos kofoidi* Chatton (Gymnodiniales) *Phycologia, 19*, 239–236.

NIE, D. (1943). Dinoflagellata of the Hainan region. IV. On the genus *Diplopsalis. Sinensia, 14*, 1–21.

NORRIS, D.R. & BERNER, L.D. (1970). Thecal morphology of selected species of *Dinophysis* (Dinoflagellata) from the Gulf of Mexico. *Contr. in Marine Science, 15*, 145–92.

NORRIS, R.E. (1967). "Micro-algae in enrichment cultures from Puerto Penasco, Sonora, Mexico". *Bull. S. Calif. Acad. Sci., 66*, 233–50.

OSTENFELD, C.H. (1901). Phytoplankton fra det Kaspiske Hav. *Vidensk. Medd. dansk. naturh. Foren. Kbh.*, 129–39.

OSTENFIELD, C.H. (1908). The phytoplankton of the Aral Sea, and its affluents, with an enumeration of the Algae observed. *Wiss. Ergebnisse der Aralsee. Exp. Lief., 8*, 128–225.

PARKE, M. & BALLANTINE, D. (1957). A new marine dinoflagellate : *Exuviaella mariae-lebouriae* n. sp. *J. mar. biol. Ass. U.K., 36*, 643–50.

PARKE, M. & DIXON, P.S. (1976). Check-list of British marine algae – third revision. *J. mar. biol. Ass. U.K., 56*, 527–94.

PAULSEN, O. (1908). XVIII Peridiniales. In *Nordisches Plankton.* Ed. K. Brandt & C. Apstein. Keil und Leipzig.

PAULSEN, O. (1912). Peridiniales Ceterae. *Bull. Trim. Bur. Cons. Perm. Int. Expl. Mer. Copenhagen,* 251–90.

PAULSEN, O. (1931). Etudes sur le microplancton de la mer d'Alboran. *Trabajos del Instituto Espanol de Oceanografia, 4*, 1–108.

PAULSEN, O. (1949). Observations on Dinoflagellates. *K. Danske Vidensk. Selsk. Skr., 6* (4), 1–67.

PAVILLARD, J. (1913). Le Genre *Diplopsalis* Bergh et les Genres voisins. Montpellier, 1–12.

PAVILLARD, J. (1931). Phytoplancton (diatomees, peridiniens) proverant des campagnes scientifiques du Prince Albert I de Monaco. *Result. Camp. Sci. Monaco, 82*, 1–208.

PENNICK, N.C. & CLARKE, K.J. (1977). The occurrence of scales in the peridinian dinoflagellate *Heterocapsa triquetra* (Ehrenb.) Stein. *Br. phycol. J., 12*, 63–66.

PETERS, N. (1930). Peridinea. Die Tierwelt der Nord- und Ostsee. Leipzig, Part II, 13–84.

PINGREE, R.D., HOLLIGAN, P.M. & MARDELL, G.T. (1978). The effects of vertical stability on phytoplankton distributions in the summer on the northwest European shelf. *Deep Sea Res., 25*, 1011–28.

PINGREE, R.D., HOLLIGAN, P.M., MARDELL, G.T. & HEAD, R.N. (1976). The Influence of physical stability on spring, summer and autumn phytoplankton blooms in the Celtic Sea. *J. mar. biol. Ass. U.K., 56*, 845–73.

PINGREE, R.D., MADDOCK, L. & BUTLER, E.I. (1977). The influence of biological activity and physical stability in determining the chemical distributions of inorganic phosphate, silicate and nitrate. *J. mar. biol. Ass. U.K., 57*, 1065–73.

POUCHET, G. (1883). Contribution a l'histoire des Cilio-Flagelles. *J. Anat. Physiol. Paris, 19*, 399–455.

POUCHET, G. (1885). Contribution a l'histoire des Peridiniens marins. *J. Anat. Physiol. Paris, 21*, 28–88.

PRAGER, J.C. (1963). Fusion of Fam. Glenodiniaceae into Peridiniaceae, notes on *Glenodinium foliaceum* Stein. *J. Protozool., 10*, 195–204.

REID, P.C. (1972). Dinoflagellate cyst distribution around the British Isles. *J. mar. biol. Ass. U.K., 52*, 939–44.

REID, P.C. (1974). Gonyaulacacean dinoflagellate cysts from the British Isles. *Nova Hedwigia, 25*, 579–630.

REID, P.C. (1977). Peridiniacean and Glenodiniacean dinoflagellate cysts from the British Isles. *Nova Hedwigia, 29*, 429–456.

REID, P.C. (1978). Dinoflagellate cysts in the plankton. *New Phytol., 80*, 219–29.

REINECKE, P. (1967). *Gonyaulax grindleyi* sp. nov. A dinoflagellate causing red tide at Elands Bay, Cape Province, in December 1966. *J.S. Afr. Bot., 33*, 157–60.

SCHILLER, J. (1928). Die planktischen Vegetationen des adriatischen Meeres. C. Dinoflagellata. 1 Teil. Adiniferidea, Dinophysidaceae. *Arch. Protistenk., 61*, 45–91.

SCHILLER, J. (1933). Dinoflagellatae. Vol. 10, III, pt. 1. In *Rabenhorsts' Kryptogamenflora*. Leipzig.

SCHILLER, J. (1937). Dinoflagellatae. Vol. 10, III, pt. 2. In *Rabenhorsts' Kryptogamenflora*. Leipzig.

SCHMITTER, R.E. (1971). The fine structure of *Gonyaulax polyedra*, a bioluminescent marine dinoflagellate. *J. Cell Sci., 9*, 147–73.

SCHUTT, F. (1887). Uber die Sporenbildung mariner Peridineen. *Ber. dtsch. bot. Ges., 5*, 364–74.

SCHUTT, F. (1895). Die Peridineen der Plankton-Expedition. *Ergebnisse der Plankton-Expedition der Humboldt-Stiftung, 4*, 1–170.

SILVA, E. de S. (1959). Some observations on marine dinoflagellate cultures. 1. *Prorocentrum micans* and *Gyrodinium* sp. *Notas Estudos, Lisboa, 21*, 1–15.

SKUJA, H. (1939). Beitrag zur Algenflora Lettlands, II. *Acta Hortibot. Univ. Letviensis, 11–12*, 41–169.

SOLUM, I. (1962). Taxonomy of *Dinophysis* populations in Norwegian Waters in view of biometric observations. *Nytt Magasin for Botanikk, 10*, 5–32.

SOURNIA, A. (1967). Genus *Ceratium* in Mozambique Channel. A contribution to the comprehensive revision of the genus. Parts I and II. *Vie Milieu, Ser. A., 18*, 375–500.

SOURNIA, A. (1973). Catalogue des especes et taxons infraspecifiques de Dinoflagelles marins actuels. I. Dinoflagelles libres. *Beihefte Nova Hedwigia, 48*, 1–92.

SOURNIA, A., CACHON, J. & CACHON, M. (1975). Catalogue des especes et taxons infraspecifiques de Dinoflagelles marins actuels publies depuis la revision de J. Schiller. II. Dinoflagelles parasites ou symbiotiques. *Arch. Protistenk, 117*, 1–19.

STEIDINGER, K.A. (1964). *Gymnodinium splendens* Lebour. *Fla. Bd. Cons. Mar. Lab. Leaf. Ser. 1*, No. 3.

STEIDINGER, K.A. (1968). The genus *Gonyaulax* in Florida waters. I. Morphology and thecal development in *Gonyaulax polygramma* Stein. *Fla. Bd. Cons. Mar. Lab. Leaf. Ser. 1*, No. 4.

STEIDINGER, K.A. (1973). Phytoplankton ecology : a conceptual review based on eastern Gulf of Mexico research. *CRC Critical Reviews in Microbiol., 3*, 49–68.

STEIDINGER, K.A. (1979). Collection, enumeration and identification of free-living marine dinoflagellates. In *Toxic Dinoflagellate Blooms*. Ed. D.L. Taylor & H.H. Seliger. Elsevier, North Holland. pp 435–42

STEIDINGER, K.A. & DAVIS, J.T. (1967). The genus *Pyrophacus* with a description of a new form. *Fla. Bd. Cons. Mar. Lab.Leaf. Ser. 1*, No. 3, 1–8

STEIDINGER, K.A. & TRUBY, E.W. (1978). Ultrastructure of the red tide dinoflagellate *Gymnodinium breve*. General description *J. Phycol., 14*, 72–79.

STEIDINGER, K.A. & WILLIAMS, J. (1970). Memoirs of the Hourglass Cruises. II. Dinoflagellates. *Publs. Mar. Res. Lab., Fla. Dept. Nat. Res., 2*, 1–251.

STEIN, F. (1883). *Der Organismus der Infusionstiere*. III. pt. 2. Die Naturgeschichte der Arthrodelen Flagellaten. Leipzig. pp. 1–30.

STOSCH, H.A. von (1964). Zum Problem der sexuellen fortpflanzung in der Peridineengattung *Ceratium. Helgol. Wiss, Meeresunters., 10*, 140–52.

STOSCH, H.A. von (1969a). Dinoflagellaten aus der Nordsee. 1. Uber *Cachonina niei* Loeblich (1968), *Gonyaulax grindleyi* Reinecke (1967) und eine Methode zur Darstellung von Peridineen panzern. *Helgol. Wiss. Meeresunters., 19*, 558–68.

STOSCH, H.A. von (1969b). Dinoflagellaten aus der Nordsee. 2. *Helgolandinium subglobosum* gen. et sp. nov. *Helgol. Wiss. Meeresunters., 19*, 569–77.

STOSCH, H.A. (1973). Observations on vegetative reproduction and sexual cycles of two freshwater dinoflagellates *Gymnodinium pseudopalustre* Schiller and *Woloszynskia apiculata* sp. nov. *Br. Phycol. J., 8*, 105–34.

SUBRAHMANYAN, R. (1966). New spp. of Dinophyceae from Indian waters. I. The genera *Haplodinium* Klebs emend. Subrahmanyan, and *Mesoporos* Lillick. *Phykos, 5*, 175–80.

SWEENEY, B.M. (1971). Laboratory studies of a green *Noctiluca* from New Guinea. *J. Phycol., 7*, 53–58.

SWIFT, E. & DURBIN, E.G. (1971). Similarities in the asexual reproduction of the oceanic dinoflagellates *Pyrocystis fusiformis. P. lunula* and *P. noctiluca. J. Phycol., 7*, 89–96.

SWIFT, E. & WALL, D. (1972). Asexual reproduction through a thecate stage in *Pyrocystis acuta* Kofoid, 1907 (Dinophyceae). *Phycologia, 11*, 57–65.

TAI, Li-Sun & SKOGSBERG, T. (1934). Studies on the Dinophysidae, Marine Armoured Dinoflagellates, of Monterey Bay, California. *Arch. Protistenk., 82*, 380–482.

TANGEN, K. (1977). Blooms of *Gyrodinium aureolum* in North European waters, accompanied by mortality in marine organisms. *Sarsia, 63*, 123–33.

TAYLOR, D.L. (1971a). Taxonomy of some common *Amphidinium* Species. *Br. phycol. J., 6*, 129–33.

TAYLOR, D.L. (1971b). Ultrastructure of the Zooxanthella *Endodinium chattonii* in situ. *J. mar biol. Ass. U.K., 51*, 227–34.

TAYLOR, D.L. (1971c). On the symbiosis between *Amphidinium klebsii* and *Amphiscolops langerhansi* (Turbellaria:Acoela). *J. mar. biol. Ass. U.K., 51*, 301–14.

TAYLOR, F.J.R. (1962). *Gonyaulax polygramma* Stein in Cape waters: a taxonomic problem related to developmental morphology. *J. S. Afric. Bot., 28*, 237–42.

TAYLOR, F.J.R. (1972). Unpublished observations on the thecate stage of the dinoflagellate genus *Pyrocystis* by the late C.A. Kofoid and Josephine Michener. *Phycologia, 11*, 47–55.

TAYLOR, F.J.R. (1975a). Non-helical transverse flagella in Dinoflagellates. *Phycologia, 14*, 45–47.

TAYLOR, F.J.R. (1975b). Taxonomic difficulties in red tide and paralytic shellfish poison studies: the 'Tamarensis complex' of *Gonyaulax. Environ Lett., 9*, 103–19.

TAYLOR, F.J.R. (1976). Dinoflagellates from the International Indian Ocean Expedition. *Biblthca bot., 132*, 1–234.

TAYLOR, F.J.R. (1979). The toxigenic gonyaulacoid dinoflagellates. 47–56. In *Toxic Dinoflagellate Blooms*. Ed. D.L. Taylor & H.H. Seliger.

TETT, P.B. (1971). The relation between dinoflagellates and the bioluminescence of sea water. *J. mar. biol. Ass. U.K., 51*, 183–206.

TOMAS, R.N. & COX, E.R. (1973). Observations on the symbiosis of *Peridinium balticum* and its intracellular alga. I. Untrastructure *J. Phycol., 9*, 304–23.

TOMAS, R.N., COX, E.R. & STEIDINGER, K.A. (1973). *Peridinium balticum* (Levander) Lemmermann, an unusual dinoflagellate with a mesocaryotic and an eucaryotic nucleus. *J. Phycol., 9*, 91–98.

TURPIN, D.H., DOBELL, P.E.R. & TAYLOR, F.J.R. (1978). Sexuality and cyst formation in Pacific strains of the toxic dinoflagellate *Gonyaulax tamarensis. J. Phycol., 14*, 235–38.

TUTTLE, R.C. & LOEBLICH, A.R. III (1975). Sexual reproduction and segregation analysis in the dinoflagellate *Crypthecodinium cohnii. J. Phycol. suppl., 11*, 15.

WALL, D. (1965). Modern Hystrichospheres and dinoflagellate cysts from the Woods Hole Region. *Grana Palynologica, 6*, 297–314.

WALL, D. & DALE, B. (1968a). Modern dinoflagellate cysts and evolution of the Peridiniales. *Micropaleontology, 14*, 265–304.

WALL, D. & DALE, B. (1968b). Quaternary Calcareous Dinoflagellates (Calciodinellideae) and their Natural Affinities. *J. Paleont., 42*, 1395–1408.

WALL, D. & DALE, B. (1971). A reconsideration of living fossil *Pyrophacus* Stein, 1883 (Dinophyceae). *J. Phycol.,* **7**, 221–35.

WOLOSZYNSKA, J. (1928). Dinoflagellates der polnischen Ostsee sowie der an Piasnica gelegenen Sumpfe. *Arch. Hydrobiol. Rybact. 3*, 153–278.

YARRANTON, G.A. (1967). Parameters for use in distinguishing populations of *Euceratium* Gran. *Bull. mar. Ecol., 6*, 147–59.

ZINGMARK, R.G. (1970). Sexual reproduction in the dinoflagellate *Noctiluca miliaris* Suriray. *J. Phycol., 6*, 122–26.

ADDENDUM

ANON (1973). Continuous plankton records: A plankton atlas of the North Atlantic and the North Sea. *Bull. mar. Ecol.* **7**, 1–174.

BOALCH, G.T. (1979). The dinoflagellate bloom on the coast of South-west England, August-September 1978. *J. mar. biol. Ass. U.K. 59*, 515–528.

DODGE, J.D. (1975b). The Prorocentrales (Dinophyceae). II. Revision of the genus *Prorocentrum. Bot. J. Linn. Soc. 71*, 103–125.

DODGE, J.D. & HERMES, H. (1981). A revision of the *Diplopsalis* group of dinoflagellates (Dinophyceae) based on material from the British Isles. *Bot. J. Linn. Soc. 83.*

HOLLIGAN, P.M., MADDOCK, L. & DODGE, J.D. (1980). The distribution of dinoflagellates around the British Isles in July 1977: a multi-variate analysis. *J. mar. biol. Ass. U.K. 60*, 851–867.

OTTWAY, B., PARKER, M., MCGRATH, D & CROWLEY, M. (1979). Observations on a bloom of *Gyrodinium aureolum* Hulburt on the south coast of Ireland, Summer 1976, associated with mortalities of littoral and sub-littoral organisms. *Irish Fish Invest.* Ser. B No. *18* 1–9.

PYBUS, C. (1980). Observation on a *Gyrodinium aureolum* (Dinophyta) bloom off the south or Ireland. *J. mar. biol. Ass. U.K. 60*, 661–674.

THRONDSEN, J. (1969). Flagellates of Norwegian coastal waters. *Nytt. mag. Bot. 16*, 161–216.

WALL, D. & EVITT, W.R. (1975). A comparison of the modern genus *Ceratium* Schrank, 1973, with certain Cretaceous marine dino-flagellates. *Micropaleontology 21*, 14–44.

INDEX TO GENERA AND SPECIES

Names used in the monograph are in roman type. The genera described are in capital letters with the main page indicated by bold type. Synonyms are in italics.

Printed in England for Her Majesty's Stationery Office by
Hobbs the Printers of Southampton
(2568) Dd716860 C5 3/82 G381